SOCIETY FOR EXPERIMENTAL BIOLOGY
*SEMINAR SERIES: 39*

T0275768

PLANTS UNDER STRESS

# SOCIETY FOR EXPERIMENTAL BIOLOGY SEMINAR SERIES

A series of multi-author volumes developed from seminars held by the Society for Experimental Biology. Each volume serves not only as an introductory review of a specific topic, but also introduces the reader to experimental evidence to support the theories and principles discussed, and points the way to new research.

# PLANTS UNDER STRESS

*Biochemistry, Physiology and Ecology*
*and their Application to Plant Improvement*

*Edited by*

## HAMLYN G. JONES

*Head of Crop Production Division*
*Institute of Horticultural Research, Wellesbourne*

## T.J. FLOWERS

*Reader in Plant Physiology*
*School of Biology, University of Sussex*

## M.B. JONES

*Lecturer in Botany and Fellow,*
*Trinity College, University of Dublin*

CAMBRIDGE
UNIVERSITY PRESS

CAMBRIDGE UNIVERSITY PRESS
Cambridge, New York, Melbourne, Madrid, Cape Town, Singapore, São Paulo

Cambridge University Press
The Edinburgh Building, Cambridge CB2 8RU, UK

Published in the United States of America by Cambridge University Press, New York

www.cambridge.org
Information on this title: www.cambridge.org/9780521344234

First published 1989
Reprinted 1992, 1993
This digitally printed version 2008

*A catalogue record for this publication is available from the British Library*

*Library of Congress Cataloguing in Publication data*
Plants under stress/edited by Hamlyn G. Jones, T. J. Flowers,
M. B. Jones.
p. cm. – (Society for Experimental Biology seminar series: 39)
Includes index.
ISBN 0 521 34423 9
1. Crops – Effect of stress on – Congresses.
2. Plants, Effect of stress on – Congresses.
3. Plants breeding – Congresses.
I. Jones, Hamlyn G.
II. Flowers, T. J. (Timothy J.)
III. Jones, M. B. (Michael B.) 1946–
IV. Series: Seminar series (Society for Experimental Biology
(Great Britain)); 39.
SB112.5.P54   1989
632 – dcl9   88-36689   CIP

ISBN 978-0-521-34423-4 hardback
ISBN 978-0-521-05037-1 paperback

# CONTENTS

# CONTRIBUTORS

*Austin, R.B.*
Institute of Plant Science Research (Cambridge Laboratory), Cambridge CB2 2LQ.

*Blum, A.*
Institute of Field and Garden Crops, The Volcani Centre, Agricultural Research Organisation, POB 6, Bet Dagan, Israel.

*Bryant, J. A.*
Department of Biological Sciences, University of Exeter, Exeter EX4 4QG.

*Davies, W.J.*
Division of Biological Sciences, University of Lancaster, Lancaster LA1 4YQ.

*Evans, J.R.*
Plant Environmental Biology Group, Research School of Biological Sciences, Australian National University, GPO Box 475, Canberra City, ACT 2601, Australia.

*Farquhar, G.D.*
Plant Environmental Biology Group, Research School of Biological Sciences, Australian National University, GPO Box 475, Canberra City, ACT 2601, Australia.

*Flowers, T.J.*
School of Biological Sciences, The University of Sussex, Brighton BN1 9QG.

*Grime, J.P.*
Unit of Comparative Plant Ecology (NERC), Department of Plant Sciences, The University, Sheffield S10 2TN.

*Ho, T.D.*
Department of Biology, Washington University, St Louis, Missouri 63130, USA.

*Hubick, K.T.*
Plant Environmental Biology Group, Research School of Biological Sciences, Australian National University, GPO Box 475, Canberra City, ACT 2601, Australia.

*Hughes, S.G.*
Nuovo Crai, Ricerca Centrale del Gruppo S.M.E., 'La Fagianeria' 81015
Piana di Monte Verna (CE), Italy.

*Jones, H.G.*
Institute of Horticultural Research, Wellesbourne, Warwick CV35 9EF.

*Jones, M.B.*
School of Botany, Trinity College, University of Dublin, Dublin, Eire.

*Pritchard, J.*
Centre for Arid Zone Studies and School of Biological Science, University
College of North Wales, Bangor, Gwynedd, Wales LL57 2UW.

*Rains, D.W.*
Department of Agronomy and Range Science, University of California,
Davis, California 95616, USA.

*Sachs, M.M.*
Department of Biology, Washington University, St Louis, Missouri 63130,
USA.

*Sharp, R.E.*
Department of Agronomy, University of Missouri, Columbia, Missouri
65211, USA.

*Smirnoff, N.*
Department of Biological Sciences, University of Exeter, Exeter EX4
4QG.

*Stewart, G.R.*
Department of Biology (Darwin Building), University College London,
Gower Street, London WC1E 6BT.

*Wong, S.C.*
Plant Environmental Biology Group, Research School of Biological
Sciences, Australian National University, GPO Box 475, Canberra City,
ACT 2601, Australia.

*Woodward, F.I.*
Department of Botany, University of Cambridge, Downing Street, Cambridge CB2 3EA

*Wyn Jones, R. G.*
Centre for Arid Zone Studies and School of Biological Science, University
College of North Wales, Bangor, Gwynedd, Wales LL57 2UW.

*Yeo, A.R.*
School of Biological Sciences, The University of Sussex, Brighton BN1
9QG.

# PREFACE

The study of plant responses to stress has been a central feature of plant biologists' attempts to understand how plants function in their natural and managed (agricultural/horticultural) environments. In recent years, much progress has been made in understanding how stresses affect plant performance and consequently we considered it timely to examine the ecological, physiological and biochemical responses of plants to a variety of stresses with a view to identifying common principles. In particular we were interested to see if we are now in a position to evaluate ways in which this information might be applied by plant breeders, particularly through some of the newer methods of genetic engineering, to the development of more stress-resistant crop plants.

The chapters in this book were contributed by the invited speakers at a two and a half day meeting organised by the Environmental Physiology Group of the Society for Experimental Biology and held during the Society's annual conference at Lancaster in March 1988. The meeting was also supported by the Association of Applied Biologists and the British Ecological Society. The book is divided into three parts. The first is an introductory section (Chapters 1–3) which outlines essential concepts and considers the relative importance of different stresses as limiting factors for plant production. The second section (Chapters 4–7) concerns the behaviour of plants under stress at different levels of organisation, while the third section (Chapters 8–13) considers the application of this information on stress to crop improvement both through conventional breeding and molecular biology.

We hope this book will be of particular interest to senior undergraduates and research students seeking an introduction to the area of plant stress physiology but we also trust that it will be of interest to a wide range of plant scientists including molecular biologists, physiologists, agriculturalists, ecologists and plant breeders concerned with stressful environments.

We would like to take this opportunity to thank for their help those concerned with organising the meeting at Lancaster University, particularly the local secretaries Dr W.J. Davies and Professor P.J. Lea, as well as Dr L.D. Incoll the convenor and the committee of the Environmental Physiology Group. We also gratefully acknowledge the financial support of the Association of Applied Biologists and the British Ecological Society and last but not least we thank the contributors for producing their manuscripts so promptly.

H.G. Jones (*Wellesbourne*)
T.J. Flowers (*Sussex*)
M.B. Jones (*Dublin*)

H.G. JONES AND M.B. JONES

# 1 Introduction: some terminology and common mechanisms

## Introduction

The study of plant responses to stress has been a central feature of environmental physiologists' attempts to understand how plants function in their natural environment and in particular to explain patterns of plant distribution and their performance along environmental gradients (Osmond *et al.*, 1987). The best of these studies have had a vertical integration from cellular to ecosystem processes (e.g. Björkman, 1981). The primary object of this volume is to bring together contributions from ecologists, physiologists, molecular biologists and plant breeders, who each have their own perspectives on stressful environments and how plants perform in them, and to consider how their understanding of plant responses may be applied by plant breeders. Unfortunately in a book of this size it is not possible to cover all types of stress. In particular, detailed treatment of plant responses to temperature extremes has been omitted as it is the subject of a recent SEB symposium volume (Long & Woodward, 1988), and the extensive potentially relevant work on, for example, pollution, pests and pathogens, and mineral nutrition has been omitted.

In recent years the extended controversy concerning the appropriate terminology to use in studies of plant responses to stressful environments (e.g. Kramer, 1980; Levitt, 1980; Harper, 1982) has often detracted attention from the identification and understanding of underlying principles. Despite this it is useful at this stage to outline the main concepts involved and attempt to provide a generally acceptable common framework for further discussions.

## What is stress?

The term *stress*, when used in biology, has general connotations rather than a precise definition (Osmond *et al.*, 1987). It is therefore most useful to apply the term stress in its more general sense as an 'overpowering pressure of some adverse force or influence' (*Shorter Oxford English Dictionary*, 3rd edn, 1983) that tends to inhibit normal systems from functioning. In general the use of the term stress should be restricted to the description of the causal influences rather than the responses.

In many cases, where one is concerned with the effects of specific environmental factors it is appropriate to replace the general term stress by the appropriate quantitative measure (e.g. soil water content or water potential) together with an appropriate measure of the plant response (e.g. growth rate).

A common approach in plant biology at the ecosystem and whole plant level that conforms with the above recommendation is to consider as stressful any situation where the external constraints limit the rate of dry-matter production of all or part of the vegetation below its 'genetic potential' (Grime, 1979). Of course, even a small difference in growth rate can have a dramatic effect on a plant's growth and competitive ability. We can already see the difficulty in being precise about stress as the response in terms of yield will depend on the severity of the stress, the time over which it is imposed, and whether or not the plant can fully recover from the effects. The external constraints or form of stress may be biotic (e.g. pests or diseases) or they may be physical and related to shortages or excesses in the supply of solar energy, water or mineral nutrients or atmospheric pollutants. For agricultural systems this simple concept probably needs modifying to consider economic yield rather than dry-matter production, though these two quantities will often be closely related. There are also difficulties in quantifying the 'genetic potential' of plants and therefore measuring the impact of stress. However, we may be able to predict dry-matter production or yield in an 'optimal' environment using a crop-growth model (e.g. Monteith, 1977) and then assess the effect of a particular stress as the reduction below the optimum.

Plant species or varieties differ in terms of their optimal environments and their susceptibility to particular stresses. Some workers prefer to consider as stressful only those environments that actually damage the plants and cause a qualitative change such as membrane damage or cell death, while others consider that in stressed systems, energy expenditure is increased or potential energy of the system is decreased (Lugo & McCormick, 1981). Both concepts seem too restrictive because mechanisms that enable plants to grow successfully in what might be expected to be sub-optimal or otherwise stressful environments could be regarded as stress-tolerance mechanisms, and would certainly be of interest to plant breeders, even though no detrimental effect is necessarily observed.

Another view (Odum, Finn & Franz, 1979) is that stress should be used to describe unfavourable deflections from the usual environment (with 'subsidy' being used to describe favourable deviations from normal). This also does not seem to be particularly useful in studies of general mechanisms of adaptation to stress, as normally extreme environments such as deserts or arctic regions with low primary productivity would not be considered stressful. Many other definitions of stress have been proposed, especially in the zoological literature, but many of these, such as Adenylate Energy Charge, a biochemical index, actually attempt to define stress in terms of the response to or impact of the environment (Ivanovici & Wiebe, 1981), so should probably be avoided.

A further difficulty is that stress factors do not usually operate alone so that interactions between and covariation of stresses are the norm in the natural environment. Stress may also have a greater effect during certain phases of the plant's life cycle than others. Seedling establishment and floral development are often particularly sensitive. Also, the complexity of biological responses means that it is often difficult to disentangle cause and effect. For example a lack of soil water is an environmental stress that can cause a water deficit in certain cells of a plant. This cell water deficit may itself be regarded as a stress that then affects various metabolic processes. This is why the term water stress is widely used although it does not conform with the favoured definition of stress given above. Similarly for salinity or low temperature stresses: they may each give rise to cellular water deficits which may be the immediate cause of many of the observed symptoms. It may therefore be appropriate to consider different factors as stresses at different levels of organisation.

We should also acknowledge the proposal that environmental stress terminology can succeed only if it is based on the principles of mechanics (Levitt, 1972, 1980). This approach is adopted by Woodward (Chapter 2) in an attempt to develop a general description of ecosystem responses to adverse environmental conditions. Woodward also develops the concept that the occurrence of rare extreme events is particularly important for ecosystem functioning. Unfortunately, attempts to impose such a restrictive definition based on the principles of mechanics can lead to confusion because these principles are not necessarily appropriate for biological systems where the chain of events arising from a given environmental factor can be complex and may involve a wide range of physical, chemical and biological processes. Although the approach may be eminently suited for the treatment of wind effects (Chapter 2) the analogy tends to break down when extended to other stresses and to biochemical responses. The ways in which non-mechanical stresses influence physiological functioning are often only loosely analogous to the stress–strain relationships of mechanics, although elastic and plastic strain may be considered as being equivalent to reversible and irreversible responses. It appears unlikely that it will be possible to derive a common unit to quantify different stresses.

### Adaptation, tolerance and resistance

The description and quantification of the plant's *response* to stress is also fraught with difficulties. Wherever possible it may be preferable to follow Woodward (1987) and avoid the use of emotive terminologies such as 'strategy' or 'tactics' because many workers consider their use in the discussion of plant responses as philosophically objectionable as these words are said to be teleological (e.g. Kramer, 1980; Harper, 1982). Nevertheless, strategy is a useful word that does not necessarily have these implications and that can validly be used with due care to describe a genetically programmed series of responses that can favour survival of that genotype in a given environment. This is not teleological.

There are particular problems with 'adaptation' as it is used in two senses: firstly with respect to the contribution of a character to the fitness of an organism to survive in its present environment and secondly with respect to the evolutionary origin of a character. In this book we are primarily concerned with the former more general use, though the application to crop improvement requires that the characters are heritable. In the second sense, therefore, it is worth noting that the process of adaptation involves a heritable modification in structure or function. This contrasts with 'acclimation' which occurs during the life of an organism and is not heritable. 'Hardening' is often used to mean acclimation. An illustration of the difference between adaptation and acclimation is given by the genetic adaptation of photosynthesis in *Eucalyptus* to temperature at various elevations, on which is superimposed short-term acclimation to seasonal variation in temperature (see Kramer, 1980). Note that the results of both adaptation and acclimation processes can be adaptive (in the first sense).

The quantification of adaptation is difficult because it is unlikely that any plant is in a state of perfect adaptation to its environment since it is made up of a collection of ancestral characteristics and the process of adaptation is occurring continually. Indeed, Harper (1982) has argued that we should refer to abaptation rather than adaptation – evolution from rather than evolution towards. We can say that adaptation to an environment depends on the possession of an optimum combination of characters that minimises deleterious effects and maximises advantageous effects (Bradshaw, 1965). We must bear in mind, however, that non-adaptive characters may evolve in parallel with adaptive characters by pleiotropy, and that the direction of adaptive change is limited by the available genetic resources of the species (Harper, 1982). This is part of the reason why Harper (1982) argues that those features that we observe in an organism in nature are not necessarily optimal solutions to past selective forces.

The use of tolerance and resistance also poses problems. Although we favour the use of tolerance as the general term describing all mechanisms whereby plants may in some way perform better in a stressful environment (that is, we follow Kramer (1980) rather than Levitt (1972, 1980)), resistance is still preferred by some workers (Chapter 11). It seems to us that resistance is somewhat misleading, implying a rather passive response, while tolerance is much the more widely used term for some stresses such as salinity. Nevertheless it can be useful to have two different words available to distinguish particular situations, in which case it is essential that they are clearly defined when first used. It is, of course, possible to subdivide apparent stress tolerance into categories such as 'Escape' (where the plant avoids being subject to the stress), 'Avoidance' (where the plant avoids its tissues being subject to the stress even though the stress is present in the environment) and true 'Tolerance' at a biochemical or physiological level (e.g. Jones, 1983).

### Are there common features of different stresses and plant response?

There are several aspects of different environmental stresses that either have common features or the plant responses or adaptations to those stresses may have common components or indicate general principles. It is an objective of this volume to identify such features where they exist so as to help in the development of stress-tolerant crop plants by making the best use of the newer techniques of molecular biology. Particular examples will be discussed in more detail in succeeding chapters.

To predict the impact of stress on plants we need to know something about (i) the temporal variation in stresses, (ii) the plant's potential to acclimatise to stress and (iii) interactions between different stresses and the plant responses. As pointed out earlier, stresses can be very disparate in nature, even though some such as water deficits, freezing and salinity may act, at least partly, through a common mechanism. In other cases different stresses often occur together, for example high temperature and drought, or waterlogging and mineral toxicities. Furthermore different stresses can interact, both in their occurrence and in their effects, so that real progress in crop improvement is likely to depend on an improved understanding of these interactions and their consequences for plants and ecosystems. Because of this complexity it may not be useful to pursue too far the theme of common features of different environmental stresses.

On the other hand, there is substantial evidence that plant responses to apparently very different stresses can have important common features. A particular theme that recurs throughout this volume (especially in Chapters 5, 6, 10, 11 and 12) is the importance of osmotic adjustment and related phenomena such as changes in cell wall rheology. Alterations in the cellular osmotica, whether maintaining turgor or protecting membranes, enzymes or organelles from damage, appear to be crucial in adaptation to drought and salinity and also confer some protection to temperature extremes. Whether osmoregulation is also relevant for other stresses remains to be established.

Another topic of wide interest is the altered synthesis of proteins when plants or other organisms are subjected to a range of environmental stresses (Chapters 8, 9 and 10). Many of these 'stress proteins' are highly conserved across a wide range of organisms and in some cases similar proteins can be induced by very different stresses. The proteins produced in response to high temperature stress are particularly well characterised, but there is an increasing body of information on proteins induced by other stress such as salinity, anoxia, heavy metals, UV light and pathogenesis. Although some proteins are induced by a range of stresses (e.g. UV light and fungal elicitors both cause synthesis of some enzymes involved in phenlyalanine metabolism), in most cases distinct proteins are synthesised in response to different stresses, and there is only limited evidence for common stress proteins. The extent to which common proteins reflect common features of cell damage, rather than being involved in stress tolerance, remains to be determined. The investigation of the role of stress proteins in stress tolerance will be an important area of study in the future.

Some other examples of common plant responses that are not discussed in great detail elsewhere in this volume, but that are nevertheless worth highlighting include:

*(i) Cross protection.* It is well known that exposure to certain stresses can cause a degree of hardening such that plants become more tolerant not only of that stress, but also of other stresses. This is particularly the case for drought and frost, and for salinity and frost (e.g. Schmidt *et al.*, 1986) or chilling (Rikin, Blumenfeld & Richmond, 1976), and implies some common feature in the tolerance mechanisms. In such cases the common feature may be related to the common involvement of cellular water deficits. In some cases the cross protection that is observed may involve reduced growth and enhanced senescence and leaf abscission. Such changes tend to favour survival at the expense of overall productivity.

Cross protection is well known in plant pathology, where inoculation with one virus, for example, can give a highly specific protection against subsequent infections by that virus (Sequiera, 1983). More interestingly in the present context, inoculation of some plants, especially tobacco and some Cucurbitaceae, with virulent pathogens can induce systemic resistance to subsequent inoculations not only with the same organism, but also with a wide range of different organisms. This phenomenon is generally known as induced resistance. In addition, it is sometimes possible to induce such resistance by application of salt solutions or ethylene (Métraux & Boller, 1986), by heat shock (Stermer & Hammerschmidt, 1984), or even by stem injections of β-ionone (a compound related to abscisic acid) (Salt, Tuzun & Kuć, 1986). Much remains to be done in elucidating the mechanisms of these effects, and although a range of processes has been implicated, it is possible that the characteristic pathogenesis-related proteins may be involved, since some of the proteins induced by an initial infection (or by physical or chemical inducers) have been shown to have chitinase activity (Métraux & Boller, 1986).

Another phenomenon that may be relevant here is the systemic induction of proteinase inhibitor synthesis in undamaged leaves when other parts of the plant are wounded (see Ryan *et al.*, 1985). It is thought that this reaction represents part of a defence strategy against insect attack. The detailed mechanism of the signalling involved is not known, although oligosaccharins have been implicated (Ryan *et al.*, 1985). It may even have something in common with the induction of systemic resistance.

*(ii) Abscisic acid.* Most interest in the endogenous plant growth regulator abscisic acid (ABA) has centred on its possible role in drought tolerance, such that several groups have put significant effort into screening varieties for differences in ABA production in response to drought (see Chapters 11 & 13). There have been several suggestions that ABA acts as a 'stress hormone' involved in plant acclimation to a wide range of stresses, including the development of induced resistance, of both a localised and possibly even a systemic nature. Certainly it has been invoked as a factor in plant adaptive responses to drought, frost, high temperature, waterlogging and salinity (see Jones, 1979, 1983), and there are indications that it may be involved in a wider range of stress responses including attack by pests and pathogens. As an example of the latter effect, infection of tobacco by tobacco mosaic virus (TMV) can cause a fourfold increase of ABA levels in the leaves (Whenham *et al.*, 1986), with the most marked increases in ABA occurring within dark green areas in the leaves where TMV multiplication is less rapid. Taken together with the observation that exogenous ABA at low concentrations can significantly decrease susceptibility to infection by TMV (Fraser, 1982), and the response to β-ionone (mentioned above), these results imply a role for ABA in the control of resistance to infection. It may

be particularly relevant to note that adaptive responses often involve developmental changes, which may be partially mediated by alterations in abscisic acid levels.

*(iii) Polyamines.* In many respects the role of polyamines in plant functioning is still mysterious after many years' work. They are almost certainly involved in the control of growth and development through their interactions with nucleic acids and membranes (Smith, 1985). There is increasing circumstantial evidence for their involvement, especially of putrescine, in plant responses to a wide range of stresses including pH, Mg deficiency, osmotic shock, cold, $SO_2$ pollution, and cadmium and ammonium toxicity (Smith, 1985). It remains to be determined, however, how, and indeed whether, putrescine accumulation in response to these diverse stresses is beneficial.

### Conclusions

Although evolutionary forces have played a major role in generating the existing stress-tolerance mechanisms, the requirements of agriculture, with high yields having a higher premium than survival mechanisms, indicate that new combinations of characters are required even though they may not have evolved naturally.

Ecological and whole-plant studies have an essential role in extending the information obtained from more detailed physiological and biochemical studies so as to identify likely characters for incorporation into plant breeding programmes. It has not usually been found to be possible to extrapolate directly from physiological results, because of the complexity of the whole-plant and crop or canopy responses. For example, although abscisic acid is produced when plants are subjected to water deficits, it does not necessarily follow that a high production of abscisic acid improves the performance of a crop in drought. In fact it now appears (Quarrie, 1981) that a high production of abscisic acid in response to drought is advantageous in some species (e.g. maize and sorghum) but is actually detrimental in others (e.g. wheat). Some other examples of this type of problem are highlighted by Blum (Chapter 11), Yeo & Flowers (Chapter 12) and Austin (Chapter 13).

The techniques of molecular biology have particular potential for rapidly introducing small numbers of single genes. Unfortunately there is strong evidence that the complex compensation mechanisms that exist in plants, and the interactions between different whole-plant and biochemical responses to stress, will make the direct improvement of environmental stress tolerance in crop plants by genetic engineering rather more difficult than in some other areas such as pathogen tolerance (Chapter 8). Nevertheless, the modern techniques of molecular biology are already providing a powerful new tool for increasing our understanding of physiological processes involved in plant response to, and adaptation to, environmental stresses (Chapters 8, 9 and 13).

## References

Björkman, O. (1981). The response of photosynthesis to temperature. In *Plants and their Atmospheric Environment*, ed. J. Grace, E.D. Ford and P.G. Jarvis, pp. 273–301. Oxford: Blackwell Scientific Publications.

Bradshaw, A.D. (1965). Evolutionary significance of phenotypic plasticity in plants. *Advances in Genetics*, **13**, 115–55.

Fraser, R.S.S. (1982). Are 'pathogenesis-related' proteins involved in acquired systemic resistance of tobacco plants to tobacco mosaic virus? *Journal of General Virology*, **58**, 305–13.

Grime, J.P. (1979). *Plant Strategies and Vegetation Processes*. Chichester: John Wiley.

Harper, J.L. (1982). After description. In *The Plant Community as a Working Mechanism*, Special Publication No. 1 of the British Ecological Society, ed. E.I. Newman, pp. 11–25. Oxford: Blackwell Scientific Publications.

Ivanovici, A.M. & Wiebe, W.J. (1981). Towards a working 'definition' of 'stress': A review and critique. In *Stress Effects on Natural Ecosystems*, ed. G.W. Barrett & R. Rosenberg, pp. 13–27. New York: Wiley.

Jones, H.G. (1979). PGRs and water relations. In *Aspects and Prospects of Plant Growth Regulators*, ed. B. Jeffcoat, pp. 91–9. Letcombe: British Plant Growth Regulator Group.

Jones, H.G. (1983). *Plants and Microclimate*. Cambridge: Cambridge University Press.

Kramer, P.J. (1980). Drought stress and adaptation. In *Adaptation of Plants to Water and High Temperature Stress*, ed. N.C. Turner & P.J. Kramer, pp. 6–20. New York: Wiley.

Levitt, J. (1972). *Responses of Plants to Environmental Stresses*, 1st edn. New York: Academic Press.

Levitt, J. (1980). Stress terminology. In *Adaptation of Plants to Water and High Temperature Stress*, ed. N.C. Turner & P.J. Kramer, pp. 437–43. New York: Wiley.

Long, S.P. & Woodward, F.I. (1988). *Plants and Temperature*. Oxford: Blackwell Scientific Publications (in press).

Lugo, A.E. & McCormick, J.F. (1981). Influence of environmental stressors upon energy flow in a natural terrestrial ecosystem. In *Stress Effects on Natural Ecosystems*, ed. G.W. Barrett & R. Rosenberg, pp. 79–102. New York: Wiley.

Métraux, J.P. & Boller, Th. (1986). Local and systemic induction of chitinase in cucumber plants in response to viral, bacterial and fungal infections. *Physiological and Molecular Plant Pathology*, **28**, 161–9.

Monteith, J.L. (1977). Climate and the efficiency of crop production in Britain. *Philosophical Transactions of the Royal Society of London B*, **281**, 277–94.

Odum, E.P., Finn, J.T. & Franz, E. (1979). Perturbation theory and the subsidy–stress gradient. *BioScience*, **29**, 349–52.

Osmond, C.B., Austin, M.P., Berry, J.A., Billings, W.D., Boyer, J.S., Dacy, J.W.H., Nobel, P.S., Smith, S.D. & Winner, W.E. (1987). Stress physiology and the distribution of plants. *BioScience*, **37**, 38–48.

Quarrie, S.A. (1981). Genetic variability and heritability of drought-induced abscisic acid accumulation in spring wheat. *Plant, Cell & Environment*, **4**, 147–51.

Rikin, A., Blumenfeld, A. & Richmond, A.E. (1976). Chilling resistance as affected by stressing environments and abscisic acid. *Botanical Gazette*, **137**, 307–12.

Ryan, C.A., Bishop, P.D., Walker-Simmons, M., Brown, W. & Graham, J. (1985). The role of pectic fragments of the plant cell wall in response to biological stresses. In *Cellular and Molecular Biology of Plant Stress*, ed. J.L. Key & T. Kosuge, pp. 319–34. New York: Alan R. Liss.

Salt, S.D., Tuzun, S. & Kuć, J. (1986). Effects of β-ionone and abscisic acid on the growth and resistance to blue mold. Mimicry of effects of stem infection by *Peronospora tabacina* Adam. *Physiological and Molecular Plant Pathology*, **28**, 287–97.

Schmidt, J.E., Schmitt, J.M., Kaiser, W.M. & Hincha, D.K. (1986). Salt treatment induces frost hardiness in leaves and isolated thylakoids from spinach. *Planta*, **168**, 50–5.

Sequiera, L. (1983). Mechanisms of induced resistance in plants. *Annual Review of Microbiology*, **37**, 51–79.

Smith, T.A. (1985). Polyamines. *Annual Review of Plant Physiology*, **36**, 117–43.

Stermer, B.A. & Hammerschmidt, R. (1984). Heat shock induces resistance to *Cladosporium cucumerinum* and enhances peroxidase activity in cucumbers. *Physiological Plant Pathology*, **25**, 239–49.

Whenham, R.J., Fraser, R.S.S., Brown, L.P. & Payne, J.A. (1986). Tobacco-mosaic-virus-induced increase in abscisic-acid concentration in tobacco leaves: Intracellular location in light and dark-green areas, and relationship to symptom development. *Planta*, **168**, 592–8.

Woodward, I. (1987). *Climate and Plant Distribution*. Cambridge: Cambridge University Press.

F.I. WOODWARD

# 2 The impact of environmental stresses on ecosystems

## Introduction

The use of stress terminology has been discussed in Chapter 1, where it was pointed out that the value of the term stress in indicating some adverse force or influence lies in its extreme generality, without the need for a precise quantification. Nevertheless it is appropriate that a scientific discipline should be concerned with definable quantities. This will be the starting point for this paper, which will follow the example of Levitt (1972) who applied the concepts and terminology of mechanical stress (force per unit area) and strain (a definable dimension change) to the study of plant responses to the environment. This approach will be developed here in an attempt to incorporate the philosophies behind stress effects into a general treatment of the responses of ecosystems to adverse environmental conditions.

## Mechanical stress

The mechanical concepts of stress are outlined in Fig. 1, with the axes reversed from that employed by mechanical engineers. The three salient features of a stress–strain response curve are shown in Fig. 1a. Initial increases in stress cause small strains but beyond a threshold, the yield stress, increasing stress causes ever increasing strains until the ultimate stress, at which point fracture occurs. The concept of the yield stress is more clearly realised when material is subjected to a stress and then relaxed to zero stress (Fig. 1b). In this case a strain is developed but is reversed perfectly – elastically – to zero strain at zero stress. In contrast, when the applied stress exceeds the yield stress (Fig. 1c) and the stress relaxes to zero, the strain does not return to zero. The material has irreversibly – plastically – extended. The extent of this plastic strain defines the residual strain.

The two features of elasticity and plasticity are the key features of stress–strain responses. The responses are dynamic and they may be totally or incompletely reversible, totally irreversible and also sensitive to various stress pretreatments, such as hardening. Defining and predicting the deviation from perfect elasticity is the central problem of materials science and will be adopted here with equal emphasis.

### Wind in forested ecosystems

Wind-throw and wind-snap of forest trees is a normal process in natural forests (Spurr & Barnes, 1980). Runkle (1982) has estimated that, annually, as much as 1% of the area of a forested ecosystem may be blown (or fall) over and create canopy gaps. Such a figure appears to be appropriate for a range of forested ecosystems from equatorial to boreal. The canopy gaps are critical for the establishment of new species and maintaining species diversity in the forest (Woodward, 1987).

Whereas gap creation by wind maintains diversity in natural forest

Fig. 1. Graphical description of stress.

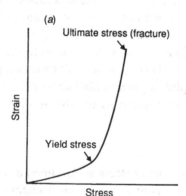

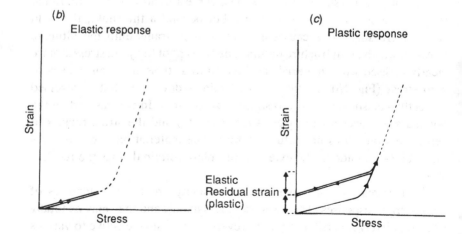

ecosystems and so may be considered desirable from a conservational standpoint, this is not the case for managed plantation forests. In contrast to natural forests with a wide age and physical structure, large areas of plantation forests are even-aged monocultures. Extremes of wind may therefore devastate whole areas if the stress of the applied wind reaches the ultimate stress point, a point which will be similar for all trees in a monoculture. This type of catastrophe was observed in the great gale in October 1987, when large areas of forest were devastated in the south-east of the British Isles. Many apparently natural woodlands were also devastated, but it should not be forgotten that many of these do not have natural origins.

The aim for tree breeders and forest managers is to define and grow a plantation which will be elastic in its response to the large stresses induced by high wind speeds. Petty & Swain (1985) have established models of the stress–strain responses of forest trees which may be used to define the sizes and morphologies of trees, for a defined range of wind speeds and elastic responses. A typical response of a plantation grown spruce tree to wind speed is shown on Fig. 2. This is a classic stress/strain curve, with an

Fig. 2. Sitka spruce sensitivity to wind (after Petty & Swain, 1985).

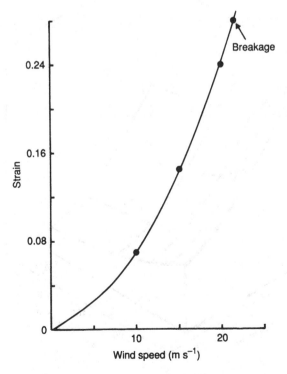

ultimate stress, or point of breakage, occurring at a wind speed of 22 m s⁻¹. At the point of breakage, the maximum resistive bending moment of the tree is exceeded. A precise definition of the applied stress, at this point, is complex because of a complex interplay of tree size, wind-speed gusting and the natural resonance of the tree.

Petty & Swain (1985) also investigated the effects of tree form, or morphology, on the susceptibility of plantations to wind-throw. This susceptibility is strongly dependent on tree taper (tree height divided by the diameter at breast height) and on tree height (Fig. 3). Trees of height 20 m will be very sensitive to only moderate wind speeds and so it is interesting to note that Sitka spruce is not typically grown beyond a maximum height of 20 m.

Trees of low taper (high values of height/DBH, where DBH is the diameter of the trunk at breast height) in the range of 120 to 140 are characteristic of trees in dense, unthinned plantations and are sensitive to only moderate wind speeds. At the other extreme, trees of high taper, with

Fig. 3. The modelled relationship between breakage wind speed, height and taper (tree height divided by the diameter at breast height) in Sitka spruce (after Petty & Swain, 1985).

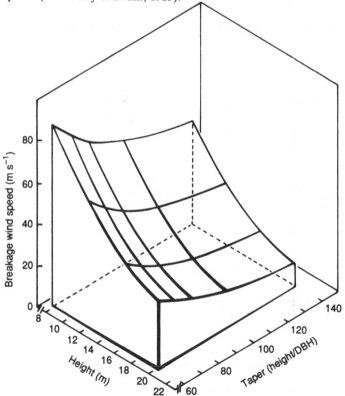

values of 60 to 70, are more resistant to high wind speeds, and are characteristic of trees in very heavily thinned plantations.

The proportion of tree weight in the photosynthetic crown also exerts an influence on the sensitivity of spruce to wind (Fig. 4). Crown weight is measured as a fraction of stem weight and clearly the smaller the crown weight fraction, the more stable the tree. However, the final requirement in terms of a sufficiently elastic tree morphology for a particular region must also consider the growth potential of the tree. For example, it is likely that a smaller crown will have a lower photosynthetic capacity than a larger crown.

This model of Petty & Swain (1985) is advanced and is clearly a valuable tool for tree breeders and managers. However, this model is essentially based on the individual tree and still needs to account for community and ecosystem properties such as canopy streamlining by leaf and twig flexing (Cionco, 1972). In addition, considerations of bud susceptibilities to

Fig. 4. The effects of varying crown weights on the sensitivity of Sitka spruce to wind speed (after Petty & Swain, 1985).

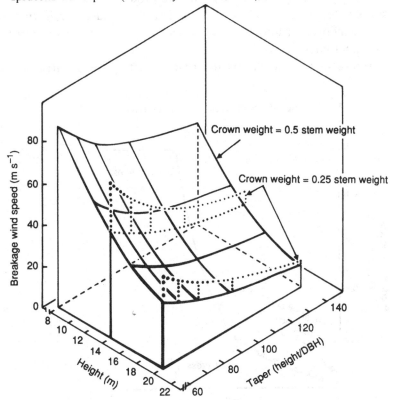

mechanical damage by twigs and branches (Jones & Harper, 1987) may also prove to be a productive avenue for exploring the development of crown morphology within tree canopies.

### The effects of mechanical stress on productivity

The strain resulting from a particular stress is formally quantified by a dimensional change, often in the length of the material under stress. Strain may be readily converted into a measure of the reduction of plant productivity, a more generally applicable strain for agriculture and ecology.

The response of productivity to stress (Fig. 5) has the same form as the strain response (Fig. 1) and emphasises the major concern of agriculture and ecology in defining and (usually) reducing the plastic residual strain, the permanent productivity reduction.

The focus on productivity in growing systems requires a time component in the study of ecosystem responses. The response of productivity to stress must therefore be considered in three dimensions (Fig. 6). This figure illustrates the effects of a stress at any particular time on the classic sigmoid curve of growth (productivity). Positive production will occur only if the stress is less than the ultimate stress and the residual strain (permanent productivity reduction) will be seen as a lowering of the growth curve below the upper boundary (the $z$ dimension in Fig. 6).

Moore & Osgood (1985) describe the effects of a hurricane on the

Fig. 5. The effects of stress on productivity.

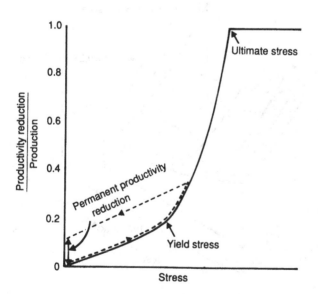

economic yield of about 5000 ha of sugar cane in Hawaii. The crop was hit by hurricane Iwa in November 1982, close to the middle of the sugar-cane cropping system of 24 months. Although the hurricane represents an extreme stress, it had a very small residual effect on the economic yield of the sugar cane, which is stored in wind-resistant stems. In contrast the whole leaf canopy of the sugar cane, which exerts the greatest friction on the momentum of the wind but contains no significant economic production, was totally stripped. In addition the young expanding stem internodes of the canopy were also broken. So the majority of the economic yield of sugar, at the time of the hurricane, was retained intact and could be harvested at the normal time (survival productivity in Fig. 6), a total of about 25% of the final harvest.

The subsequent accumulation of economic production was severely retarded until new tillers grew and started to store sugar. At harvest this new growth totalled 55% of the final yield, with a total productivity loss, resulting from the hurricane, of 20%.

Fig. 6. The effects of a hurricane on the economic yield of sugar cane (after Moore & Osgood, 1985).

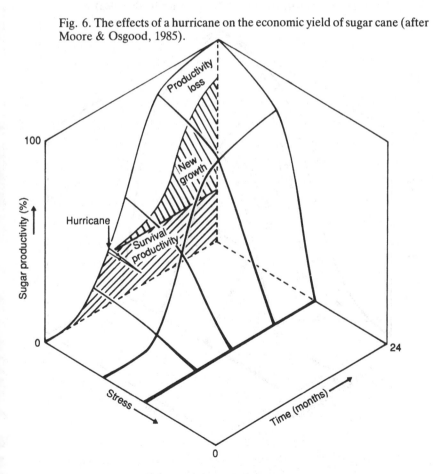

This agricultural ecosystem clearly suffered a permanent productivity reduction in response to the hurricane, but a reduction which was amplified with time and resulted from a loss of photosynthetic material. However the system was sufficiently resilient to recover from the applied stress.

### The effects of environmental extremes on ecosystems

The considerations of the effects of wind on structural damage in ecosystems conforms with the strict mechanical ideas about stress and strain. Such an approach can be expanded to consider the effects of extreme environmental conditions on ecosystem responses. An emphasis on the term extremes encourages the use of exact quantities and circumvents the use of the term stress.

The effects of an extreme winter temperature of $-10.4\,°C$ on the

Fig. 7. The sensitivity of economic yield in citrus varieties to a winter extreme low temperature of $-10.4\,°C$ (after Ikeda, 1982).

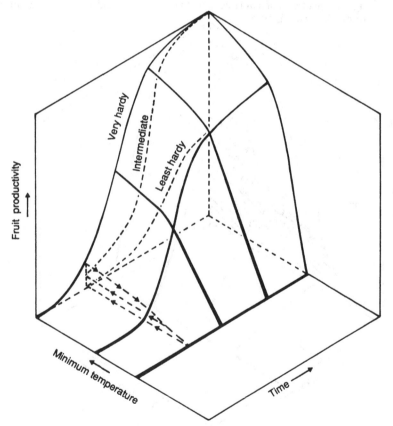

economic productivity of a range of citrus varieties in Japan, has been documented by Ikeda (1982). Four major classes of responses were observed by the different varieties (Fig. 7). The very hardy varieties suffered no obvious reductions in economic yield, with no leaf fall of these winter green crops.

Varieties which were intermediate in hardiness either had no significant yield depressions, or had a lighter crop than usual. In these varieties there was significant leaf death which will have delayed development in the growing season.

The least hardy fruit producers showed marked reductions in fruit production and significant leaf losses. In addition up to 20% of the trees were killed outright by the frost. The susceptible populations were either killed outright (up to 82%) or failed to produce any fruit crops and were generally defoliated.

Just like the sensitivity of sugar cane to the hurricane, the sensitive varieties of citrus show a certain degree of permanent productivity reduction which is seen at a maximum at harvest.

### Genetic variation in frost resistance

The obvious genetic variation in frost resistance by the varieties of citrus provides a basis for selecting genotypes to match a particular range of environments. The problem for the grower is knowing the environment, in particular the likely incidence of extreme events, such as killing frosts. There may also be an economic price to pay for plant resistance. For example it seems the most obvious approach to grow the most resistant variety which is available. However, observations in agricultural eco-systems often indicate that increased resistance is correlated with decreased yield (Li & Sakai, 1982).

A particularly clear example of this correlation is shown for different populations of *Abies sachalinensis* within a boreal forest ecosystem (Fig. 8a). Within the ecosystem different populations show an increased resistance to low temperature with altitude (Eiga & Sakai, 1984), as assessed by testing populations to frosts of either − 40 or − 43 °C. The range of genetic variability in this resistance decreases with altitude. This reduction indicates the significant limit, for breeders and for natural selection, in the range of genetic variability found within a species. Further increases by mutation will be very rare and so for applied or economic purposes further variability may only be injected by the modern biotechno-logical techniques of gene transfer.

When populations of *Abies* differing in frost resistance and altitude of origin are grown at the same sites they show marked differences in seed

Fig. 8. Frost resistance in populations of *Abies sachalinensis* within the boreal forest ecosystem. (*a*) Frost resistance and variance in resistance with altitude; (*b*) seed characteristics of populations all grown in the lowlands (after Eiga & Sakai, 1984).

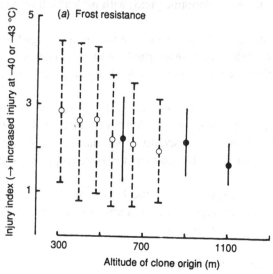

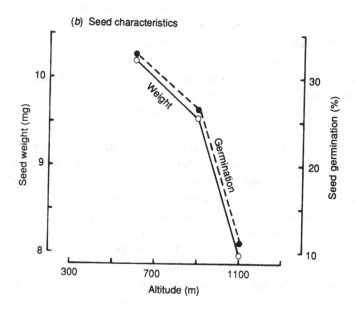

weight and capacity for germination (Fig. 8*b*). One cost of the increased frost resistance is a marked reduction in reproductive potential.

### Ecosystem recovery from environmental extremes

The examples which illustrated the responses of agricultural ecosystems to extreme events have been limited to short-term effects on productivity. For natural ecosystems which can not be replanted the recovery response to an environmental extreme is crucial, not only in the time taken for recovery but also in terms of the manner in which the ecosystem may change during and after the period of recovery. In this context, change is particularly concerned with the species which constitute an ecosystem, a consideration which is central to the aims of conservation.

A simple initial example illustrates the nature of ecosystem recovery. The response of an arctic lichen ecosystem to a short-term (six-day) field fumigation by $SO_2$ (Moser, Nash & Olafsen, 1983) indicates that recovery, measured by the recovery of photosynthetic rate, is slow (Fig. 9) and dependent on the $SO_2$ concentration during fumigation. $SO_2$ concentrations of 0.6 ppm and greater were clearly lethal. Lower doses (0.07–0.2 ppm) all caused reductions in productivity which were detectable for at least two years after fumigation. Recovery is clearly possible in the low ranges of $SO_2$ dosage, with the rate of recovery decreasing with dose.

Extreme climatic events, such as low temperatures (Woodward, 1987), may selectively exterminate particularly sensitive species within an ecosystem. However, the loss of one species in a multi-species array may have little effect on ecosystem physiognomy, e.g. the ecosystem may still

Fig. 9. Effect of $SO_2$ fumigation on productivity and subsequent photosynthetic recovery of an arctic lichen ecosystem. Fumigation occurred in 1978 for a six-day period (after Moser *et al.*, 1983).

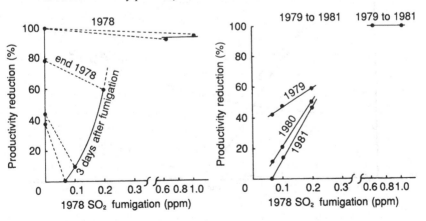

remain an evergreen forest, even though an evergreen species of tree is killed. Therefore at the ecosystem level, which covers large spatial scales, the response to an extreme may have little long-term effect. However, at the community and population levels the effect will be considerable. This effect may be charted in the response of a North American, deciduous forest ecosystem to an epidemic of a fungal pathogen, *Endothia parasitica*

Fig. 10. The effects of chestnut blight (*Endothia parasitica*) on a deciduous forest ecosystem (after Day & Monk, 1974).

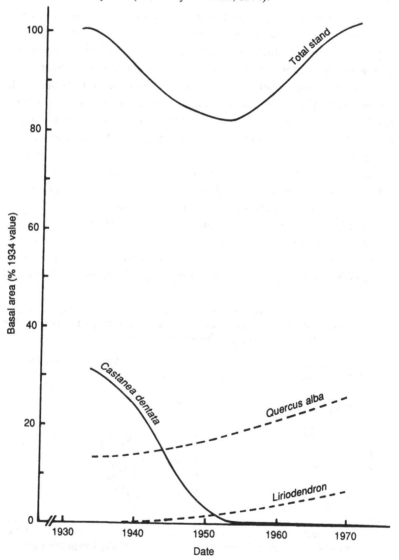

(chestnut blight). This pathogen was probably introduced to North America in the early 20th century from Asia (Woods & Shanks, 1959). The pathogen is host specific and rapidly kills even mature trees of the forest tree *Castanea dentata*, chestnut.

The fate of areas of south-eastern American deciduous forests which were dominated, before the arrival of the pathogen, by chestnut and species of oak has been traced (Fig. 10) over nearly 40 years (Day & Monk, 1974). The demise of chestnut from its standing as a canopy dominant was rapid, taking about 20 years, with a complete failure to return. During this period ecosystem productivity was reduced (as shown by a reduction in tree basal area, Fig. 10). However, the place of chestnut in the forest has been taken by a number of species, in particular oak and tulip tree, which have steadily increased in dominance. The community has clearly changed markedly but the ecosystem as a whole has shown complete recovery, in terms of re-establishing tree basal area. The place of chestnut in the oak–chestnut forests has been taken by other native species. So in the context of conservation, although the loss of chestnut is unfortunate, it has not been at the cost of an invasion of alien species of tree.

Such ecosystem resilience is not, however, universal. Marked and irreversible effects of man-induced overgrazing can occur in ecosystems, particularly in natural grasslands which are periodically droughted. Overgrazing of arid grasslands leads to a low cover of vegetation and an increased susceptibility of the top soil to erosion, by rainstorms and by wind (Cloudsley-Thompson, 1979). Once soil erosion has occurred extensively the possibility of natural recovery becomes increasingly remote. The most rapid reversal can be instituted by man, in the same way that he was the agent of the initial decline. Reversal is enhanced by a combination of seeding and marked reductions in grazing (Bridge *et al.*, 1983). The problems of selecting a species which will colonise successfully are severe. Cox *et al.* (1983) found that over a period of 92 years of tests in North America, only 5% of a suite of 300 tested species would establish, and then only in one of two or three years. Of these 5% of species, 79% were aliens.

In the grasslands of southern Arizona the high grazing pressures have exerted a marked effect on the species composition of the native grassland. Morton (1985) describes the demise of a whole phenological class of grasses in this grassland ecosystem. Before the periods of overgrazing in early 20th century the grassland contained a mixture of warm-season and cool-season grasses, which in turn depended on the bimodal nature of precipitation in summer and in winter. Overgrazing occurred particularly in the cool season, causing the total elimination of the cool-season grasses (Fig. 11).

The ecosystem had therefore been modified to a summer rain dependent prairie, with limited presence of vegetation in the winter.

Overgrazing was reduced in the 1970s and the ecosystem productivity recovered rapidly over a nine-year period but the majority of this increase was due to an invading, alien species *Eragrostis lehmanniana* (Lehmann lovegrass). This is a warm-season grass but it is more cold tolerant and grows earlier and later in the year than the native warm-season grasses (Morton, 1985). As a result the species can utilise both the winter and summer rains and is consequently more productive than the native species (Martin & Morton, 1980). The greater productivity and competitiveness of *Eragrostis* has also lead to a continual decrease in the occurrence of native warm-season grasses (Fig. 11). In terms of ecosystem resilience the presence of *Eragrostis* serves to maintain a grassland ecosystem; however, if *Eragrostis* should be subject to a pathogen, in the same manner as chestnut, then a catastrophic breakdown of the ecosystem may occur.

Catastrophic breakdown of ecosystems can also be observed in natural ecosystems, without the influence of alien species. One such global problem is forest dieback (Mueller-Dombois, 1986), which is occurring over the globe without obvious cause. Hosking & Hutcheson (1986) have made observation on dieback in the evergreen broadleaved forests of New

Fig. 11. Forage production of grassland ecosystems subject to initial overgrazing (early 20th century, degree of production estimated) and subsequent invasion by the alien grass, *Eragrostis lehmanniana* (after Morton, 1985).

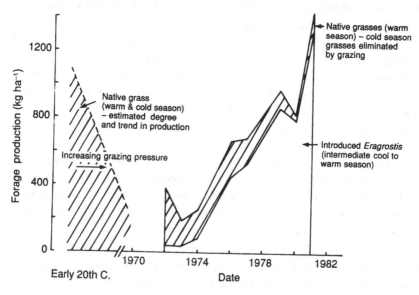

Zealand. In the North Island, at moderate altitudes (300–600 m), the hard beech (*Nothofagus truncata*) may be the sole and dominant species of the forest. Since the mid-1970s there has been a marked dieback of *Nothofagus*, without replacement and particularly at lower altitudes.

For hard beech there appears to be some evidence for the cause of the ecosystem dieback. Hosking & Hutcheson (1986) measured tree productivity at sites where the ecosystem was stable and at sites where dieback was occurring. Associated with these dynamics were differing levels of infestation by the leaf-mining weevil *Neomycta publicaris*. High infestation by the weevil causes premature leaf fall and occurs at the sites of dieback. Growth of hard beech in this area (Fig. 12*b*) has been steadily declining since the early 1970s. In addition, the absolute growth rates of these plants were only 30% of those at sites of low infestation and high beech survival.

Unfortunately the long-term nature and extent of the leaf-miner infestations are not known. What is clear is that the (likely) combination of weevil infestation in a virtual monoculture of hard beech, with a period of drought in the early 1970s, has led to a continuous decline in the growth of hard beech at the site of dieback (Fig. 12*b*). At the other site, with lower levels of weevil infestation, the plants have recovered, somewhat, after the drought (Fig. 12*a*).

### Conclusion: relative importance of stresses and extremes to ecosystem resilience

In the example of hard beech it is difficult to assess which ecological feature is causing dieback. It seems that the leaf miners predispose the plants to a marked sensitivity to drought. However, without the occurrence of the drought in the early 1970s it seems possible that dieback may not have occurred. The controls and causes of the leaf-miner infestation and severity are unknown, but are clearly a key issue for investigation.

Many effects of stress or climatic extremes are nested within a natural environmental suite of extremes. In the case of arid grasslands, the presence of grassland rather than forest is controlled by the low levels of precipitation (Woodward, 1987) and the regular incidence of drought, while a more productive ecosystem based on winter rains is likely to be limited by extremes of low temperature during the winter (Woodward, 1987). It is clear that overgrazing is secondary to these environmental limitations in that man can exert a significant impact, just by reducing the grazing pressure. Viewed in those terms it appears that the importance of extreme events relates to the ability of man to reverse their effects.

Should climate change so that extremes of low temperature are more frequent, causing the death of sensitive species, then this will not be exactly

Fig. 12. Sensitivity of growth in *Nothofagus truncata* to the combined effects of drought and leaf-miner (*Neomycta publicaris*) infestation. (*a*) Stable site with low *Neomycta* infestation; (*b*) dieback site with high *Neomycta* infestation (after Hosking & Hutcheson, 1986).

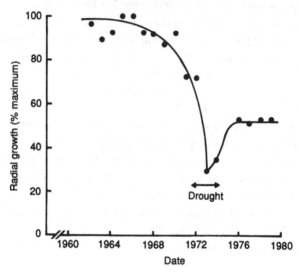

(a)   Low *Neomycta* infestation

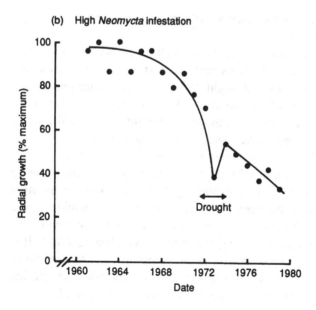

(b)   High *Neomycta* infestation

Table 1. *Hierarchy of extreme events in ecosystems*

| Events | Scale | Examples |
|---|---|---|
| Temperature extremes | Global | 1. Ice Age<br>2. Nuclear winter |
| ↓ | | |
| Water availability | Macro/Global | 1. Sahel drought |
| ↓ | | |
| Forest clearance | Macro | 1. 300 000 ha/week<br>2. Forest irreversibly to grassland |
| ↓ | | |
| Overgrazing | Macro | 1. Grassland irreversibly to desert |
| ↓ | | |
| Pollution | Macro | 1. Forest decline |
| ↓ | | |
| Fire | Meso | 1. Fire ecosystems |
| ↓ | | |
| Pathogens | Meso | 1. Chestnut blight |
| ↓ | | |
| Wind | Meso | 1. October 1987 'hurricane' in UK |

reversible. Similarly if droughts become more frequent they will also not be reversible at an ecosystem level. Such extremes will have the greatest importance for ecosystem survival. In contrast extremes such as overgrazing, pathogen infestations, wind-throw and pollution are lower in the level of hierarchy, in that man is capable of effecting significant ecosystem repair.

The extreme events which have been discussed differ in their characteristic spatial scales of importance. Each event which causes wind-throw occurs on a smaller spatial scale than a pollution event, as recently seen by the very extensive scale of the Chernobyl fallout (Johnston, 1987). At an even greater scale, the projected extent of the 'nuclear winter' will be global (Covey, 1987). Examples of this hierarchy of scales of events are shown in Table 1.

These scales of extent define the manner in which the ideas of extreme events are applicable to agriculture and forestry. Wind-throw is a small-scale process and the forest manager must grow and manage plantation forests which will survive extreme wind speeds at the spatial scale of his plantation. The question is, how rare must this extreme event be, before it may be discounted. The answer is likely to be dominated by economics and also the longevity of the plants; with a 50 year harvesting interval, it seems clear that the return periods of extreme events up to 50 years (at least) are crucial.

Should global warming continue (Jones *et al.*, 1988) then the frequency of extreme events associated with this change in mean climate will also change, but in a non-linear fashion (Wigley, 1985). One possible consequence is a marked increase in extreme events such as high temperatures. A few consecutive days of temperatures over 35 °C can markedly reduce corn yields (Wigley, 1985). So projecting corn yields in an ameliorated climate can only be reliable when the effects of extreme events are considered. Crops need to be selected for tolerance of some defined extreme event, as shown for the example of frost tolerance in citrus (Fig. 7).

### References

Bridge, B.J., Mott, J.J., Winter, W.H. & Hartigan, R.J. (1983). Improvement in soil structure resulting from sown pastures on degraded areas in the dry savanna woodlands of northern Australia. *Australian Journal of Soil Research*, **21**, 83–90.

Cionco, R.M. (1972). A wind profile index for canopy flow. *Boundary Layer Meteorology*, **3**, 255–63.

Cloudsley-Thompson, J.L. (1979). *Man and the Biology of Arid Zones*. Madison: University Park Press.

Covey, C. (1987). Protracted climatic effects of massive smoke injection into the atmosphere. *Nature*, **325**, 701–3.

Cox, J.R., Morton, H.L., La Baume, J.T. & Renard, K.G. (1983) Reviving Arizona's rangelands. *Journal of Soil Water Conservation*, **38**, 342–5.

Day, F.P. & Monk, C.D. (1974). Vegetation patterns on a southern Appalachian watershed. *Ecology*, **55**, 1064–74.

Eiga, S. & Sakai, A. (1984). Altitudinal variation in freezing resistance of Saghalien fir (*Abies sachalinensis*). *Canadian Journal of Botany*, **62**, 156–60.

Hosking, G.P. & Hutcheson, J.A. (1986). Hard beech (*Nothofagus truncata*) decline on the Mamaku Plateau, North Island, New Zealand. *New Zealand Journal of Botany*, **24**, 263–9.

Ikeda, I. (1982). Freezing injury and protection of citrus in Japan. In *Plant Cold Hardiness and Freezing Stress*, vol. 2, ed. P.H. Li & A. Sakai, pp. 575–89. London: Academic Press.

Johnston, K. (1987). British sheep still contaminated by Chernobyl fallout. *Nature*, **328**, 661.

Jones, M. & Harper, J.L. (1987). The influence of neighbours on late growth of trees. II. The fate of buds on long and short shoots in *Betula pendula*. *Proceedings of the Royal Society of London, B* **232**, 19–33.

Jones, P.D., Wigley, T.M.L., Folland, C.K., Parker, D.E., Angell, J.K., Lebedeff, S. & Hansen, J.E. (1988). Evidence for global warming in the past decade. *Nature*, **332**, 790.

Levitt, J. (1972). *Responses of Plants to Environmental Stresses*. New York: Academic Press.

Li, P.H. & Sakai, A. (1982). *Plant Cold Hardiness and Freezing Stress*, Vol. 2. London: Academic Press.

Martin, S.C. & Morton, H.L. (1980). Responses of false mesquite, native grasses and forbs and Lehmann lovegrass after spraying with picloram. *Journal of Range Management*, **33**, 104–6.

Moore, P.H. & Osgood, R.V. (1985). Assessment of sugarcane crop damage and yield loss by high winds of hurricanes. *Agricultural and Forest Meteorology*, **35**, 267–79.

Morton, H.L. (1985). Plants for conservation of soil and water in arid ecosystems. In *Plants for Arid Lands*, ed. G.E. Wickens, J.R. Goodin & D.V. Field, pp. 203–14. London: George Allen & Unwin.

Moser, T.J., Nash, T.H. III. & Olafsen, A.G. (1983). Photosynthetic recovery on arctic caribou forage lichens following a long-term field sulfur dioxide fumigation. *Canadian Journal of Botany*, **61**, 367–70.

Mueller-Dombois, D. (1986). Perspectives for an etiology of stand-level dieback. *Annual Review of Ecology and Systematics*, **17**, 221–43.

Petty, J.A. & Swain, C. (1985). Factors influencing stress breakage of conifers in high winds. *Forestry*, **58**, 75–84.

Runkle, J.R. (1982). Pattern of disturbance in some old-growth mesic forests of eastern North America. *Ecology*, **63**, 1533–46.

Spurr, S.H. & Barnes, B.V. (1980). *Forest Ecology*, 3rd edn. New York: J. Wiley & Sons.

Wigley, T.M.L. (1985). Impact of extreme events. *Nature*, **316**, 106–7.

Woods, F.W. & Shanks, R.E. (1959). Natural replacement of chestnut by other species in the Great Smokey Mountains National Park. *Ecology*, **40**, 349–61.

Woodward, F.I. (1987). *Climate and Plant Distribution*. Cambridge: Cambridge University Press.

J.P. GRIME

# 3 Whole-plant responses to stress in natural and agricultural systems

## Introduction

The initial section of this chapter will briefly review three attempts to devise predictions of plant responses to stress. The first, Liebig's Law of the Minimum, originates from agricultural research and places strong emphasis upon the identity of the stress most severely limiting plant growth in each environment. The second derives from population biology and focuses upon the ways in which plants under stress differ from each other in demographic responses. The third approach (Plant Strategy Theory) attempts to bridge the gap between ecophysiology and population biology and recognises recurrent forms of ecological and evolutionary specialisation which are frequently associated with characteristic stress responses.

After describing some of the main implications of Plant Strategy Theory for the study of stress responses, brief accounts are provided of two additional dimensions of variation in plant response to stress; these consist of 'stored growth' and resistance to mechanical stress.

## The search for a predictive model of plant responses to stress
### Liebig's Law of the Minimum and its successors

The foundations of intensive modern agriculture and particularly the manipulation of soil nutrient supply owe much to the recognition by Liebig in 1840 that plant yield could be stimulated by recognising the identity of the resource most limiting upon dry-matter production at a given place and time. Since many plant types (e.g. hydrophytes, shade plants, legumes, calcicoles, calcifuges, metallophytes, halophytes) differ in resource requirements according to their evolutionary history, the principle was rapidly accepted in agriculture and in ecology that plant distributions over space and in time are often influenced by the differential responses of plants to variation in limiting resources. In particular, the now celebrated correlations between fertiliser applications and botanical composition observed in the Park Grass Experiment (Brenchley & Warington, 1958) have strongly buttressed this conclusion.

In a recent and highly contentious projection of Liebig's Law of the Minimum (Tilman, 1982, 1988), the concept has been extended to provide a theoretical explanation for a wide range of ecological phenomena including various successional processes and the mechanisms of coexistence of plants in species-rich communities. There can be little doubt that differences between plants in their responses to particular resource shortages or toxic factors make a contribution to the mechanisms controlling plant distribution and vegetation dynamics. However, Tilman's ideas are strongly deterministic in the sense that little recognition is given to the role of vegetation disturbance, propagule dispersal and other stochastic processes known to influence plant distributions. Moreover, to a spectacular extent Tilman's predictions often appear to be at variance with empirical evidence based on studies of species distributions within communities (e.g. Hillier, 1984; Mahdi & Law, 1987; Pearce, 1987); these studies suggest an essentially random distribution of individual plants within some ancient species-rich grasslands and provide no evidence in support of Tilman's suggestion that coexistence depends upon the establishment of an equilibrium between the differing resource demands of plants and the underlying resource mosaic of the habitat.

### Plant population biology

A quite different approach to the study of plant responses to stress has been explored by those ecologists who have followed the example of Harper (1977) in applying to plants techniques originally deployed in investigations of animal populations. Here the methodology has been demographic and the resulting data have allowed responses to stress to be analysed in terms of fluctuations in the rates of mortality and recruitment of either plant populations or plant parts (e.g. leaves, inflorescences).

There can be little doubt that certain stresses have characteristic demographic consequences both for populations and for plant parts and it is also evident that, according to taxonomic and evolutionary history, species and genotypes may differ in their responses to the same stress. Thus, demographic study is often a useful preliminary to analysis of the causation and evolutionary origins of particular stress responses. However, some demographers go further and advocate long-term field observation as the only reliable way forward:

> "The detailed analysis of proximal ecological events is the only means by which we can reasonably hope to inform our guesses about the ultimate causes of the ways in which organisms behave." *Harper (1982)*

This clearly overstates the potential of demographic study to provide a mechanistic understanding of plant responses to environments and, if implemented, would lead to unnecessary delay in the development of generalising principles. The remainder of this chapter is founded on the assumption that the most direct route to a coherent predictive theory of plant responses to stress is likely to involve a synthesis of insights derived from plant population biology, ecophysiology, and many other fields of botanical endeavour.

### Plant Strategy Theory

On first inspection patterns of evolutionary and ecological specialisation in plants present a spectacle of bewildering and almost limitless complexity. Similarly the field of stress response appears to provide for numerous dimensions of variation according to the nature and intensity of stress and the many variable attributes of plants and their immediate environments. However, it has been the achievement of the pioneers of ecological theory (Macleod, 1894; Raunkiaer, 1934; Ramenskii, 1938; Hutchinson, 1959; MacArthur & Wilson, 1967; Odum, 1969) to recognise that certain paths of ecological specialisation are widely recurrent in all major taxa and in each of the world's ecosystems. These basic types (primary strategies) take different forms in the regenerative (juvenile) and established (mature) phases of the life cycle but they show consistent correlations with habitat and they are recognisable through possession of characteristic sets of traits, among which modes of stress response are prominent. Implicit in plant strategy theory is the assertion that despite numerous superficial differences ('fine-tuning') habitats impose selection forces which are predictable and fall into a small number of basic types. Also important is the notion (Grime, 1965; Stebbins, 1974) that the potentiality of the organism itself is severely constrained such that adaptive specialisation to one type of existence and habitat condition usually involves the assumption of genetic traits (including types of stress response) which preclude success in other environmental conditions.

Debate continues with regard to the number and exact nature of the primary strategies and their role in population and community dynamics and in ecosystem structure. For our present purpose, this discussion can be largely ignored; here it is necessary only to recognise that strategy theory has introduced both priorities and predictions into our thinking about the functional significance of different types of response to stress. These will now be considered in relation to one variant of plant strategy theory – the 'Competitor–Stress tolerator–Ruderal' or C–S–R model (Grime, 1974).

### Stress responses predicted by the C–S–R model

Habitats differ in the length and quality of the opportunities they provide for resource capture, growth and reproduction. Following the arguments of Southwood (1977) and Grime (1974, 1979), the matrix *habitat duration × habitat productivity* gives rise to a range of conditions and associated functional specialisation which occupies a triangular area and can form the basis for a universal classification of plants (Fig. 1*a*). Efforts to test the triangular model and explore the relevance of this model to vegetation dynamics have been described elsewhere (Grime, 1987, 1988); here attention will be confined to its use in the prediction and elucidation of plant responses to stress. This subject can be approached first by considering the link between plant strategy and two major forms of stress response – morphological plasticity and cellular acclimation – and secondly by examining the role of stress responses in mechanisms of resource capture.

Fig. 1. (*a*) Model describing the various equilibria between competition, stress and disturbance in vegetation and the location of primary and secondary strategies. C, competitor; S, stress tolerator; R, ruderal; C–R, competitive–ruderal; S–R, stress-tolerant ruderal; C–S, stress-tolerant competitor; C–S–R, 'C–S–R strategist'. ——, relative importance of competition ($I_c$); — · — ·, relative importance of stress ($I_s$); – – –, relative importance of disturbance ($I_d$). (*b*) Scheme relating pattern of seasonal change in shoot biomass to strategy. 1, competitor; 2, competitive–ruderal; 3, C–S–R strategist; 4, stress-tolerant competitor; 5, ruderal; 6, stress-tolerant ruderal; 7, stress tolerator.

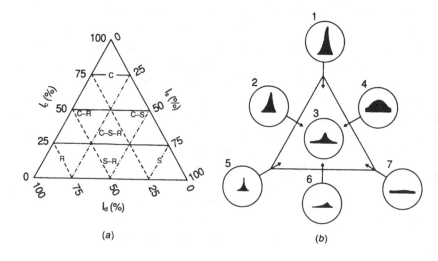

(*a*)                    (*b*)

*Morphological plasticity and cellular acclimation*

When plants are exposed to stress as a consequence of resource depletion or climatic fluctuation, many different responses are possible depending upon the species and the nature and severity of the stress. However, as Bradshaw (1965) recognised, stress responses can be classified into two basic types, one of which is morphological and the other physiological. One of the major uses of plant strategy theory is to provide a basis for predicting which of the two mechanisms is likely to be operative in particular species, populations and situations.

In order to explain the linkages between strategy and stress response reference will be made to Fig. 1*b* which depicts the patterns of seasonal change in shoot biomass associated with the full spectrum of primary strategies (Fig. 1*a*). For simplicity, this diagram refers to the patterns observed in herbaceous plants in a temperate zone situation with a sharply defined growing season. However, the principles adduced can be applied to any life-form or biome.

At position 1, corresponding to conditions of high productivity and low disturbance, the growing season is marked by the rapid accumulation of biomass above and below ground. The fast-growing perennials, e.g. *Urtica dioica*, *Phalaris arundinacea*, which usually enjoy a selective advantage under these conditions, achieve the initial surge in spring growth by mobilisation of reserves from storage organs; this quickly produces a dense and extensive leaf canopy and root network. As development continues, resources are rapidly intercepted by the plant and its neighbours and local stresses arise through the development of depletion zones which expand above and below the soil surface. In this swiftly changing scenario it is predicted that morphological plasticity, acting mainly but not exclusively at the time of cell differentiation, will be the dominant form of stress response. Attenuation of stems, petioles and roots provides a mechanism of escape from the depleted sectors of the environment and in conjunction with the short life-spans of individual leaves and roots, results in a process of continuous relocation of the effective leaf and root surfaces during the growing season. In theory, the high turnover of leaves and roots might also be expected to allow, at the stage of organ differentiation, a process of continuous modulation of leaf and root biochemistry as conditions change during the growing season.

Position 5 in Fig. 1*b* describes the brief explosive development of biomass in an ephemeral species capable of exploiting a productive but temporary habitat. Here again morphological plasticity would be expected to predominate. In the vegetative phase, plasticity in root and shoot morphology will be an integral part of the mechanism of resource capture.

However, as first elaborated by Salisbury (1942), there is a 'reproductive imperative' in the biology of ephemerals. In many species the effect of plastic response to stresses imposed by environment or crowding by neighbours is to induce early flowering and to sustain reproductive allocation as a proportion of the biomass even in circumstances of extreme stunting and early mortality. Although Salisbury's observation has been confirmed by subsequent investigations (Harper & Ogden, 1970; Hickman, 1975; Boot, Raynal & Grime, 1986), it is clear that more research is required to document precisely the stress cues which in particular ephemeral species promote early flowering. Some indication of the complications which may arise in natural habitats, where several stresses may coincide, can be gained from Fig. 2 which examines flowering allocation in a population of *Poa annua* exposed to various stresses. It is evident from these data that an unpredictable situation could arise where neighbours have depleted several resources simultaneously and conflicting signals to reproductive allocation originate from moisture stress and shading.

Under stable conditions of extremely low productivity imposed by mineral nutrient stress (position 7 in Fig. 1*b*) there is little seasonal change in biomass. Leaves and roots often have a functional life of several years and there is usually an uncoupling of resource capture from growth (Grime, 1977; Chapin, 1980). Because of the slow turnover of plant parts, differentiating cells occupy a small proportion of the biomass and morphogenetic

Fig. 2. Comparison of the effects of various stress treatments upon reproductive effort in *Poa annua*. The intensity of each stress treatment is characterised by comparing the growth increment of stressed plants to that of control (unstressed) plants over the same experimental period. □, control; ○, water stress (polyethylene glycol); ▲, mineral nutrient stress (dilute concentrations of Rorison solution); ● shading by neutral filters (Smit, 1980).

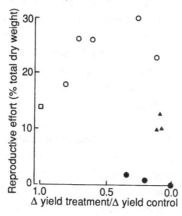

changes are less likely to provide a viable mechanism of stress response. In these circumstances, where the same tissues experience a sequence of different climatic stresses with the passing of the seasons, the dominant form of stress response is likely to be that of cellular acclimation whereby the functional characteristics and 'hardiness' of the tissues change rapidly through biochemical adjustments of membranes and organelles. These changes are reversible and can occur extremely rapidly in certain species (Hosakawa, Odani & Tagawa, 1964; Mooney & West, 1964; Strain & Chase, 1966; Björkman, 1968; Mooney, 1972; Taylor & Pearcy, 1976; Oechel & Collins, 1973; Larsen & Kershaw, 1975).

By confining attention to the three extreme contingencies of Fig. 1*b* it has been possible to emphasise the strong linkage between the stress responses of plants and the productivity and durational stability of their habitats. It is important to recognise, however, that many plants exploit habitats of intermediate status, corresponding to the central areas of the triangular model; here we may expect to find greater scope for morphogenetic plasticity and cellular acclimation to coexist within the same genotypes.

### Stress responses and resource capture

Recently studies have been conducted to test the hypothesis that plants from productive and unproductive habitats differ in morphological plasticity when exposed to resource stress. The experiments also measure the extent to which the response (or lack of response) influences resource capture under stress conditions mimicking those characteristic of productive and unproductive habitats. The designs used are analagous to some already applied to animals and can be legitimately described as resource foraging investigations. An account of some of these studies has been published (Grime, Crick & Rincon, 1986; Crick & Grime, 1987); here only the main methods and results will be summarised.

The experiments utilise partitioned containers in which each plant can be exposed to controlled patchiness in supply of light or mineral nutrients. Some treatments maintain spatially predictable patches of resources for periods long enough to allow morphogenetic adjustment of leaves and roots. Other treatments involve brief pulses of enrichment (24 h mineral nutrient 'flushes' or simulated sunflecks) localised in space and random in space and time and too short to permit significant growth adjustments. The main conclusion drawn from these experiments is that although all the species so far examined have some capacity for concentration of leaves and roots in the resource-rich sectors of a stable patchy environment, this feature is much more apparent in potentially fast-growing species of productive habitats. It has been clearly established in one experiment

38     *J.P. Grime*

(Crick & Grime, 1987) that high rates of nitrogen capture from a stable patchy rooting environment are achieved by *Agrostis stolonifera*, a fast-growing species with a small but morphologically dynamic root system (Fig. 3). Under the same conditions N-capture is much inferior in *Scirpus sylvaticus*, a slow-growing species with a massive root system, but a relatively low N-specific absorption rate and slow rate of root adjustment to mineral nutrient patchiness. These results are consistent with the hypothesis that rapid morphogenetic changes ('active foraging') are an integral part of the mechanism whereby fast-growing plants sustain high rates of resource capture in the rapidly changing resource mosaics created by actively competing plants. The experiment has also yielded evidence that the larger,

Fig. 3. Comparison of the ability of two wetland species of contrasted ecology and potential growth rates (*Agrostis stolonifera*, rapid; *Scirpus sylvaticus*, slow) to modify allocation between different sectors of the root system when exposed to a nutritional patchy rooting medium. Each plant was grown in uniform nutrient-sufficient solution culture for 10 weeks, then exposed to patchy conditions by distributing the root system equally between nine radial compartments, three of which (▨) contained full strength Rorison nutrient solution. The remaining compartments (□) were provided with 1/100 Rorison solution. The value in each compartment is the mean increment of root dry matter to the compartment over 27 days (95% confidence value in brackets). In each species compartments differing significantly (*P*<0.05) in root increment have no letters in common. Root concentration index (RCI) is calculated by dividing the mean root increment in compartments receiving full strength solution by the mean increment in compartments containing 1/100 solution (Crick & Grime, 1987).

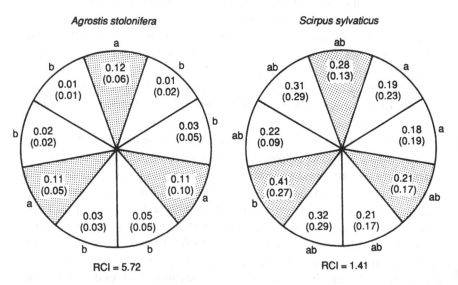

evenly distributed root system of *S. sylvaticus* enjoys an advantage when nutrients are supplied as brief localised pulses; this accords with previous studies suggesting that the major sources of mineral nutrients on infertile soils are brief episodes of nutrient mineralisation (Gupta & Rorison, 1975).

## Two further dimensions of functional specialisation and stress response

In addition to the mechanisms of stress response so far considered, there are several others which have attracted the attention of plant ecologists. These include innate or environmentally determined forms of dormancy in seeds, spores, and vegetative buds, many of which represent adaptive responses restricting plant growth and development to favourable seasons or sites. Dormancy has been the subject of numerous publications and will not be considered here. Instead, opportunity will be taken to refer to two forms of plant response to stress which until recently have received only scarce attention. The first is the phenomenon of 'stored growth' whilst the second involves the response of the developing shoot to mechanical impedance.

### Stored growth

In certain plant habitats or niches, access to resources depends crucially upon rapid growth under conditions of climatic stress. Examples of this phenomenon are particularly obvious on shallow soils in continental climates where the 'growth window' between winter cold and summer desiccation may be extremely short. In deciduous woodlands in the cool temperate zone an essentially similar niche arises in the period between snow melt and closure of the tree canopy. Both circumstances provide opportunities for high rates of photosynthesis and mineral nutrient capture in the late spring but depend upon rapid expansion of roots and shoots in the low-temperature conditions of the late winter and early spring.

Vernal geophytes provide a well-known solution to this ecological problem and controlled environment studies (e.g. Hartsema, 1961) have documented the crucial requirement for warm temperatures for the preformation of leaves and flowers during the 'inactive' summer period. Grime & Mowforth (1982) suggested that the essential feature of the vernal geophyte life cycle was the extent to which the summer developmental phase was characterised by cell division, thus circumventing the potentially limiting effect of low spring temperatures upon mitosis. According to this hypothesis, the rapid growth of the geophytes at low temperatures is mainly due to cell expansion, a process less inhibited than mitosis by low temperatures.

The geophyte growth strategy is thus associated with temporal separation of the main phases of cell division and cell expansion. It is also correlated with possession of unusually large cells (Grime & Mowforth, 1982; Grime, 1985). This suggests that another important feature of the vernal geophyte phenology is the construction of tissues which as unexpanded cells in the summer occupy a small volume in the densely packed interior of the bulb but have high expansion coefficients as they become vacuolated in the spring.

Spring growth at low temperatures is not confined to vernal geophytes, however, and it is likely that other growth-forms exhibit similar mechanisms. One source of indirect evidence here relies upon the surveys of nuclear DNA amounts now available for native plants (Bennett & Smith, 1976; Grime & Mowforth, 1982). Nuclear DNA amounts are strongly correlated with cell size and provide a convenient index of large-celled species (Olmo, 1983; Grime, 1985). This has prompted field investigations examining the relationship between the nuclear DNA amounts of species and their rates of growth in the spring. The results of these studies reveal that DNA amounts are useful both as a predictor of phenology (Fig. 4a) and as an index of temporal niche differentiation within a plant community (Fig. 4b).

Fig. 4. (a) The relationship between DNA amount and the time of shoot expansion in 24 plant species commonly found in the Sheffield region (redrawn from Grime & Mowforth, 1982). Temperature at Sheffield is expressed as the long-term average for each month of daily minima ($\square$) and maxima ($\blacksquare$) in air temperature 1.5 m above the ground. (b) The relationship between DNA amount and the mean rate of leaf extension over the period 25 March–5 April in 14 grassland species coexisting in the same turf (redrawn from Grime, Shacklock & Band, 1985). 95% confidence limits are indicated by vertical lines.

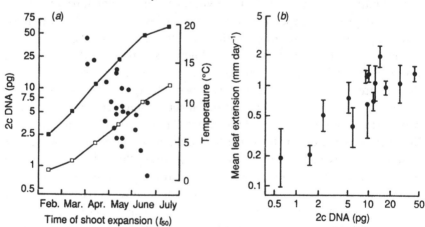

Temporal separation of cell division and cell expansion is also character-istic of the response of some grasses and cereals to moisture stress where the phenomenon has been described as 'stored growth' (Salter & Goode, 1967). Recent evidence suggests that, as in the case of the geophytes, this response is strongly dependent upon possession of large cells with high nuclear DNA amounts. Figure 5 compares the potential for stored growth in two grasses of contrasted ecology and nuclear DNA amounts by measuring their rates of leaf extension after standardised exposure to experimental drought. No capacity for stored growth is apparent in *Brachypodium pinnatum*, a species of low nuclear DNA amount and restricted in its distribution to sites with a

Fig. 5. Comparison of leaf extension rates in two grasses of contrasted cell size and nuclear DNA amount. (*a*) *Brachypodium pinnatum*, 2c DNA = 2.3 pg; (*b*) *Bromus erectus*, 2c DNA = 22.6 pg. ●, Watered prior to measurement; ○, droughted for 3 weeks prior to measurement. The 95% confidence limits are indicated by the vertical lines.

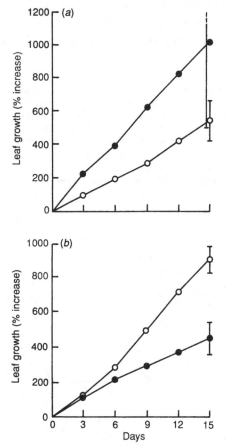

large reservoir of subsoil moisture. In marked contrast the resumption of growth after drought is characterised by a sustained surge in leaf extension in *Bromus erectus*, a species of relatively high nuclear DNA amount and distinct association with shallow soils and summer desiccation.

### Responses of expanding shoots to mechanical impedance

From field observations and experimental studies (Sydes & Grime 1981*a*, *b*), it is established that both the quantity and quality of herbaceous vegetation in deciduous woodlands is strongly affected by the varying abilities of plant species to expand their leaf canopies through the layers of tree litter which frequently accumulate on the woodland floor. It is clear, for example, that whereas the robust spear-shoot (Salisbury, 1916) of *Hyacinthoides non-scripta* can emerge from beneath a thick layer of litter, the weak mesophytic shoots of the grass *Poa trivialis* are impeded by a thin deposit of dead leaves. In view of this evidence of the differential ability of plant species to overcome mechanical resistance to shoot expansion, the possibility must be considered that the physical forces which shoots exert upon each other may be a significant factor determining the dominance hierarchies observed within closed canopies.

Fig. 6. Comparison of the responses of two grasses of contrasted growth rate and morphology to five intensities of shoot impedance. (*a*) *Lolium perenne*; (*b*) *Festuca ovina*. Each curve records the mean progress of shoot expansion in five replicate plants subjected to standardised resistances (indicated on each curve as the force in newtons required for initial deflection of weighted windows). Plants were grown individually within a transparent cone, from which the shoots, in order to escape, must deflect windows of standard dimensions and angle of inclination.

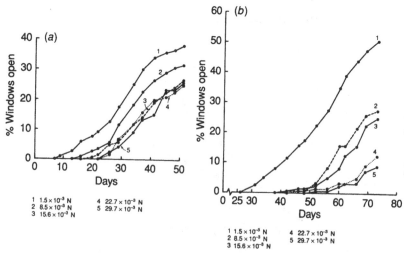

In order to test this hypothesis measurements have been made of the forces generated by the developing leaf canopies of herbaceous species (J.P. Grime, B.D. Campbell & J.C. Crick, unpublished). This has been achieved by growing each plant individually within a transparent cone, through which the shoots to escape must deflect windows of standard dimensions and angle of inclination. By altering the force required for initial displacement of the windows (by addition of lead weights) it is possible to examine the effect of increased impedence upon shoot expansion. The results, an example of which is illustrated in Fig. 6, reveal marked differences in the ability of plant species to resist physical impedance. It seems likely that mechanical stress is an important factor in the command of space and canopy dominance in herbaceous vegetation.

### Conclusions

Plant responses to stress are conditioned not only by the nature and intensity of the environmental factors involved but by the evolutionary and ecological histories of species, cultivars and genotype. Responses to resource depletion above or below ground are strongly dependent upon the productivity and durational stability of the habitat or niche to which the species or genotypes has been attuned by natural selection. Morphological plasticity is the predominant form of stress response in productive habitats; cellular acclimation is characteristic of the long-lived tissues of plants occupying habitats of consistently low productivity. Dormancy is a common mechanism of stress avoidance and is manifested in both the regenerative and established phases of plant life cycles. 'Stored growth' provides a mechanism of accelerated vegetative development under certain conditions of intermittent stress. The potential for stored growth is associated with enlargement of cell size and nuclear DNA amount. The importance of mechanical stress has been underestimated as an ecological factor. Physical interference between developing shoots may contribute to the mechanism determining the dominance hierarchy in herbaceous vegetation.

### Acknowledgements

The original research described in this paper was conducted with the support of the Natural Environment Research Council and the skilled assistance of S.R. Band and N. Ruttle. I am particularly grateful to B.D. Campbell and J.M.L. Shacklock for permission to reproduce a figure from an unpublished manuscript.

## References

Bennett, M.D. & Smith, J.B. (1976). Nuclear DNA amounts in angiosperms. *Philosophical Transactions of the Royal Society of London B*, **274**, 227–74.

Björkman, O. (1968). Carboxydismutase activity in shade-adapted and sun-adapted species of higher plants. *Physiologia Plantarum*, **21**, 1–10.

Boot, R., Raynal, D.J. & Grime, J.P. (1986). A comparative study of the influence of drought stress on flowering in *Urtica dioica* and *U. urens*. *Journal of Ecology*, **74**, 485–95.

Bradshaw, A.D. (1965). Evolutionary significance of phenotypic plasticity in plants. *Advances in Genetics*, **13**, 115–55.

Brenchley, W.E. & Warington, K. (1958). *The Park Grass Plots at Rothamsted 1856–1949*. Harpenden: Rothamsted Experimental Station.

Chapin, F.S. (1980). The mineral nutrition of wild plants. *Annual Review of Ecology and Systematics*, **11**, 233–60.

Crick, J.C. & Grime, J.P. (1987). Morphological plasticity and mineral nutrient capture in two herbaceous species of contrasted ecology. *New Phytologist*, **107**, 403–14.

Grime, J.P. (1965). Comparative experiments as a key to the ecology of flowering plants. *Ecology*, **45**, 513–15.

Grime, J.P. (1974). Vegetation classification by reference to strategies. *Nature*, **250**, 26–31.

Grime, J.P. (1977). Evidence for the existence of three primary strategies in plants and its relevance to ecological and evolutionary theory. *American Naturalist*, **111**, 1169–94.

Grime, J.P. (1979). *Plant Strategies and Vegetation Processes*. Chichester: John Wiley.

Grime, J.P. (1985). Prediction of weed and crop response to climate based upon measurements of nuclear DNA content. *Aspects of Applied Biology*, **4**, 87–98.

Grime, J.P. (1987). Dominant and subordinate components of plant communities – implications for succession, stability and diversity. In *Colonisation, Succession and Stability*, ed. A. Gray, P. Edwards and M. Crawley, pp. 413–28. Oxford: Blackwells.

Grime, J.P. (1988). The C–S–R model of primary plant strategies – origins, implications and tests. In *Plant Evolutionary Biology*, ed. L.D. Gottlieb and K. Jain, pp. 371–93. London: Chapman & Hall.

Grime, J.P. & Mowforth, M.A. (1982). Variation in genome size – an ecological interpretation. *Nature*, **299**, 151–3.

Grime, J.P., Crick, J.C. & Rincon, E. (1986). The ecological significance of plasticity. In *Plasticity in Plants*, ed. D.H. Jennings and A.J. Trewavas, pp. 5–19. Cambridge: Company of Biologists.

Grime, J.P., Shacklock, J.M.L. & Band, S.R. (1985). Nuclear DNA contents, shoot phenology and species coexistence in a limestone grassland community. *New Phytologist*, **100**, 435–45.

Gupta, P.L. & Rorison, I.H. (1975). Seasonal differences in the availability of nutrients down a podzolic profile. *Journal of Ecology*, **63**, 521–34.

Harper, J.L. (1977). *The Population Biology of Plants*. London: Academic Press.

Harper, J.L. (1982). After description. In *The Plant Community as a Working Mechanism*, Special Publication No. 1 of the British Ecological Society, ed. E.I. Newman, pp. 11–25. Oxford: Blackwell Scientific Publications.

Harper, J.L. & Ogden, J. (1970). The reproductive strategy of higher plants. I. The concept of strategy with special reference to *Senecio vulgaris* L. *Journal of Ecology*, **58**, 681–98.

Hartsema, A.M. (1961). Influence of temperature on flower formation and flowering of bulbous and tuberous plants. *Handbuch der Pflanzenphysiologie*, **16**, 123–67.

Hickman, J.C. (1975). Environmental unpredictability and plastic energy allocation strategies in the annual *Polygonum cascadense* (Polygonaceae). *Journal of Ecology*, **63**, 689–701.

Hillier, S.H. (1984). *A Quantitative Study of Gap Recolonization in Two Contrasted Limestone Grasslands*. PhD thesis, University of Sheffield.

Hosakawa, T., Odani, H. & Tagawa, H. (1964). Causality of the distribution of corticolous species in forests with special reference to the physiological approach. *Bryologist*, **67**, 396–411.

Hutchinson, G.E. (1959). Homage to Santa Rosalia or Why are there so many kinds of animals? *American Naturalist*, **93**, 145–59.

Larsen, D.W. & Kershaw, K.A. (1975). Acclimation in arctic lichens. *Nature*, **254**, 421–3.

Liebig, J. (1840). *Chemistry and its Application to Agriculture and Physiology*. London: Taylor & Walton.

MacArthur, R.H. & Wilson, E.D. (1967). *The Theory of Island Biogeography*. Princeton: Princeton University Press.

Macleod, J. (1894). Over de bevruchting der bloemen in het Kempisch gedeelte van Vlaanderen. *Deel 11. Bot. Jaarboek*, **6**, 119–511.

Mahdi, A. & Law, R. (1987). On the spatial organisation of plant species in a limestone grassland community. *Journal of Ecology*, **75**, 459–76.

Mooney, H.A. (1972). The carbon balance of plants. *Annual Review of Ecology and Systematics*, **3**, 315–46.

Mooney, H.A. & West, M. (1964). Photosynthetic acclimation of plants of diverse origin. *American Journal of Botany*, **51**, 825–7.

Odum, E.P. (1969). The strategy of ecosystem development. *Science*, **164**, 262–70.

Oechel, W.D. & Collins, N.J. (1973). Seasonal patterns of $CO_2$ exchange in bryophytes at Barrow, Alaska. In *Primary Production and Production Processes*, ed. L.C. Bliss and F.E. Wielogolaski, pp. 197–203. Stockholm: Wenner-Gren Center.

Olmo, E. (1983). Nucleotype and cell size in vertebrates: a review. *Basic and Applied Histochemistry*, **27**, 227–56.

Pearce, A.K. (1987). *An Investigation of Phenotypic Plasticity in Co-existing Populations of Four Grasses in a Calcareous Pasture in North Derbyshire*. PhD thesis, University of Sheffield.

Ramenskii, L.G. (1938). *Introduction to the Geobotanical Study of Complex Vegetations*. Moscow: Selkhozgiz.

Raunkiaer, C. (1934). *The Life Forms of Plants and Statistical Plant Geography*. Oxford: Clarendon Press.

Salisbury, E.J. (1916). The emergence of the aerial organs in woodland plants. *Journal of Ecology*, **4**, 121–218.

Salisbury, E.J. (1942). *The Reproductive Capacity of Plants*. London: Bell.

Salter, P.J. & Goode, J.E. (1967). Crop responses to water at different stages of growth. *Commonwealth Agricultural Bureau Research Review* 2. East Malling: CAB.

Smit, P.T. (1980). *Phenotypic Plasticity of Four Grass Species under Water, Light and Nutrient Stress*. BSc thesis, University of Utrecht.

Southwood, T.R.E. (1977). Habitat, the templet for ecological strategies? *Journal of Animal Ecology*, **46**, 337–65.

Stebbins, G.L. (1974). *Flowering Plants: Evolution above the Species Level*. London: Arnold.

Strain, B.R. & Chase, V.C. (1966). Effect of past and prevailing temperatures on the carbon dioxide exchange capacities of some woody desert perennials. *Ecology*, **47**, 1043–5.

Sydes, C. & Grime, J.P. (1981*a*). Effects of tree leaf litter on herbaceous vegetation in deciduous woodland. I. Field investigations. *Journal of Ecology*, **69**, 237–48.

Sydes, C. & Grime, J.P. (1981*b*). Effects of tree leaf litter on herbaceous vegetation in deciduous woodland. II. An experimental investigation. *Journal of Ecology*, **69**, 249–62.

Taylor, R.J. & Pearcy, R.W. (1976). Seasonal patterns in the $CO_2$ exchange characteristics of understorey plants from a deciduous forest. *Canadian Journal of Botany*, **54**, 1094–1103.

Tilman, D. (1982). *Resource Competition and Community Structure*. Princeton: Princeton University Press.

Tilman, D. (1988). *Plant Strategies and the Dynamics and Structure of Plant Communities*. Princeton: Princeton University Press.

G.D. FARQUHAR, S.C. WONG, J.R. EVANS
AND K.T. HUBICK

# 4 Photosynthesis and gas exchange

## Introduction

Photosynthesis and gas exchange of leaves are affected by many stresses including drought, flooding, salinity, chilling, high temperature, soil compaction and inadequate nutrition. Many, but not all, of these stresses have symptoms in common. For example, stomatal conductance and the rate of assimilation of $CO_2$ per unit leaf area often decrease when stress occurs. Further, it is possible that several of the stresses may exert their effects, in part, by increasing the levels of the hormone abscisic acid (ABA) in the leaf epidermis. This hormone is known to close stomata when applied to leaves.

There have been many studies of the effects of stress on gas exchange. Some have involved leaves that were fully expanded and some have been on leaves that were still expanding. Sometimes the stresses have been chronic, and sometimes they have been imposed for short periods. Sometimes the stresses have been severe, and sometimes they have been mild. Often they have been imposed more rapidly than occurs naturally.

It is thus easy to understand why interpretations have also been varied. There needs to be a more systematic approach in our studies, paying more attention to genetic differences that exist and including observations of what happens when stress is released. More teamwork may be needed to apply the range of techniques now at hand.

In this chapter we review recent developments, concentrating mostly on the effects of water stress because of its great importance, but trying where possible to identify general problems. As noted above, accumulation of ABA in leaves is common under stress, and we review its effects on gas exchange. If the sole direct effect of ABA were to reduce stomatal aperture, then assimilation rate would also decrease because of a lower intercellular partial pressure of $CO_2$, $p_i$. Some experiments have suggested that $p_i$ is little affected by ABA and that the *capacity* for $CO_2$ fixation has actually decreased. The situation has become complicated with the observation (Terashima *et al.*, 1988) that application of ABA can cause non-uniform stomatal closure. This has created difficulties for the interpretation of gas exchange data, and emphasised the need for alternative methods of estimating $p_i$. In species with the $C_3$ pathway of assimilation it has been shown that measurements of the discrimination against the fixation of naturally occurring $^{13}CO_2$ can be used to this end (cf. review by Farquhar *et al.*, 1988). We discuss the basis of this observation, and also how the techniques can be used to select for genotypes that have a greater ratio of $CO_2$ fixation to transpirational water loss. Finally, because it is possible that high irradiances may exacerbate the effects of stress on photosynthesis, we review the phenomenon known as photoinhibition. This term refers to light-induced reduction of photosynthetic capacity.

### Effects of water stress on gas exchange
#### Effects on stomatal conductance

The sensitivity of stomata to water stress has been well documented (see Hsiao, 1973). It was therefore accepted early that much of the reduction in the rate of $CO_2$ assimilation per unit leaf area during water stress is due to stomata impeding the inward passage of $CO_2$. Figure 1 illustrates the sequence of events which occurred during rapid imposition of water stress in a cotton plant. The experiment was conducted during mid-summer where vapour pressure deficit in the glasshouse generally exceeded 3.5 kPa. Due to the high transpiration rate, soil water potential dropped from $-0.3$ to $-2.8$ MPa in four days (day 10 to day 14 after watering was withheld). Leaf conductance, $g$, the partial pressure of $CO_2$ in the substomatal cavities, $p_i$, and in the ambient air, $p_a$, and the rate of $CO_2$ assimilation, $A$, are related as follows

$$A = g(p_a - p_i)/(1.6\ P) \tag{1}$$

where $P$ is the atmospheric pressure and the factor 1.6 arises because $g$ normally refers to the diffusion of water vapour, rather than of $CO_2$. From this relationship, it can be seen that a decrease in $p_i$ associated with a change in $g$ will cause a decrease in $A$ which is proportionally smaller than that in $g$. A convex, curved, relationship between $A$ and $g$ arises when $g$ is the sole source of variation. Such a curve was apparent when Osmond, Winter & Powles (1979) analysed data of Lawlor & Fock (1978), obtained when roots of *Zea mays* were transferred from Arnon and Hoagland solution to $-0.25$, $-0.5$ or $-1.0$ MPa polyethylene glycol solutions.

In the field, plants are normally not deprived of water rapidly. During slowly increasing water stress, the rate of $CO_2$ assimilation generally decreases in virtually the same proportion as the rate of transpiration (or

Fig. 1. Rates of $CO_2$ assimilation, A ($\mu$mol m$^{-2}$ s$^{-1}$); leaf conductance, $g$ (mol m$^{-2}$ s$^{-1}$); intercellular partial pressure of $CO_2$, $p_i$ (Pa); soil water potential and leaf water potential, $\psi$ (MPa) during gas-exchange measurements of a 30-day-old cotton plant, plotted against day after watering was withheld. Measurements were made with 2 mmol m$^{-2}$ sec$^{-1}$ photon flux density, 30 °C leaf temperature, and 2.0 kPa vapour pressure difference between leaf and air (S.C. Wong, unpublished data).

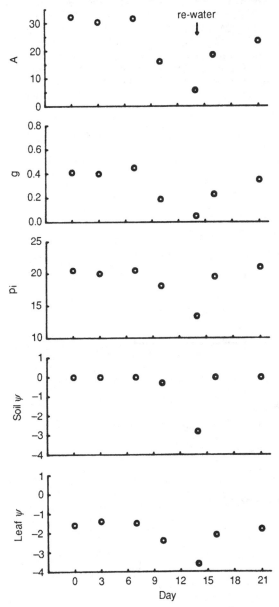

50     *G.D. Farquhar* et al.

leaf conductance). Brix (1962) showed that rates of photosynthesis and transpiration in loblolly pine and tomato decreased hand-in-hand during water stress. Rates of photosynthesis and transpiration declined in the same proportion in soybean as a result of increasing water stress (Boyer, 1970). Observations with *Acacia harpophylla* showed that $A$ and $g$ decreased in

Fig. 2. Rates of $CO_2$ assimilation, $A$, and leaf conductances, $g$, as functions of intercellular partial pressure of $CO_2$, $p_i$, in *Zea mays* on various days after withholding watering. Measurements made with 9.5, 19.0, 30.5, and 38.0 Pa ambient partial pressure of $CO_2$, 2 mmol m$^{-2}$ s$^{-1}$ photon flux density, 30 °C leaf temperature, and 2.0 kPa vapour pressure differences between leaf and air. Closed symbols represent measurements with 30.5 Pa ambient partial pressure of $CO_2$. Leaf water potentials were 0.05, $-0.2$, $-0.5$ and $-0.8$ MPa on day 0, 4, 11 and 14, respectively (after Wong *et al.*, 1985).

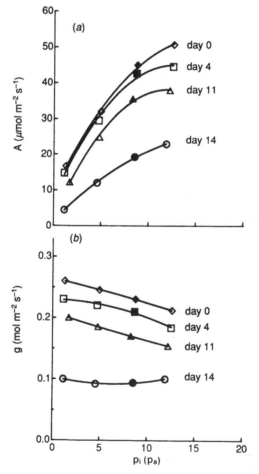

similar proportions even when phyllode water potential was less than − 5 MPa (van den Driessche, Connor & Tunstall, 1971). Takeda, Sugimoto & Agata (1978) showed that $A$ and $g$ in *Zea mays* decreased in similar proportions as leaf water content was reduced. In all the above cases there was a linear relationship, with zero intercept, between $A$ and $g$. All these authors concluded that the parallel changes between $A$ and $g$ (or rate of transpiration) during water stress were due, partly or totally, to the effect of water stress on stomata. Decrease in $A$ was assumed to be the consequence of stomatal closure. Yet this conclusion was incompatible with Equation (1). In fact, if Equation (1) is rewritten to solve for $p_i$,

$$p_i = p_a - 1.6 \, A/(Pg),$$ (2)

we see that proportional changes in $A$ and $g$ give rise to a constant value of $p_i$, so that stomatal closure could hardly have been causing the reduction of photosynthesis by reducing the availability of $CO_2$.

This point was made clearly by Wong, Cowan & Farquhar (1985). They imposed slow water stress on 30-day-old *Zea mays* plants grown in 45 l plastic bins during early spring where glasshouse vapour pressure deficit was generally about 1.0–1.5 kPa. During the 14 day period the predawn leaf water potential declined from − 0.5 to − 0.8 MPa. Rate of assimilation,

Fig. 3. Rate of $CO_2$ assimilation, $A$, and leaf conductance, $g$, in *Zea mays* on various days after withholding watering. Data are those shown in Fig. 2 with 30.5 Pa ambient partial pressure of $CO_2$ (after Wong *et al.*, 1985).

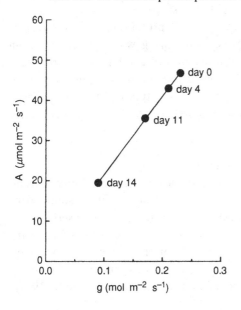

$A$, and leaf conductance, $g$, also declined (Fig. 2). At any given ambient partial pressure of $CO_2$ the relative reductions in $A$ and $g$ were in almost the same proportion (Fig. 3), so that the change in $p_i$ was not more than 1 Pa.

### Effects of water stress on biochemical capacity for photosynthesis

With such small changes in $p_i$, and yet large changes in $A$, it appears that the biochemical capacity for photosynthesis is somehow decreased by water stress. The evidence for direct short-term effects is meagre. Although Potter & Boyer (1973) reported a reduction in electron transport through photosystem II of chloroplasts isolated from rapidly desiccated (hours) Helianthus annuus leaves, Sharkey & Badger (1982) found no reduction in the uncoupled electron transport capacity of Xanthium strumarium cells, or of intact spinach chloroplasts, exposed to reduced water potential in the suspension medium. von Caemmerer & Farquhar (1984) observed that electron transport was unchanged in thylakoids isolated from water-stressed Phaseolus vulgaris leaves: maximum Rubisco activity was also unchanged. These leaves had been stressed so that $A$ decreased to zero over five days. Kaiser (1987) reviewed the subject and concluded that mesophyll photosynthetic capacity is rather insensitive to short-term dehydration down to 50–70% relative water content.

At this stage there are insufficient data on capacity after long-term stress. Powles & Björkman (1982a) found that thylakoid properties appeared unaffected by long-term water stress at low light intensities, but that primary photosynthetic reactions were damaged when the long-term stress was accompanied by natural high light intensities.

It is possible that, over these longer periods, protein synthesis is affected. Itai & Vaadia (1965, 1971) showed that during water stress cytokinin activity in the root exduate and translocation of cytokinin were reduced. Ben-Zioni, Itai & Vaadia (1967) showed that the lower rate of protein synthesis of stressed leaf tissue could be raised by treatment of leaf material with kinetin. We need more information on levels of various proteins after naturally occurring stress.

### Heterogeneity of stomatal opening

Recently queries have been raised concerning the validity of using proportional changes in $A$ and $g$ (and the maintenance of constant calculated $p_i$) as evidence that photosynthetic capacity has decreased. In the calculation of $p_i$ (Equation 2), it is assumed that $CO_2$ and $H_2O$ exchange are uniform over the area of leaf under consideration. However, it has been reported that there is a wide variation in aperture of stomata within a small area of the leaves of Oryza sativa (Ishihara, Nishihara & Ogura, 1971),

*Hordeum vulgare* L. and *Vicia faba* L. (Laisk, Oja & Kull, 1980). If patches of neighbouring stomata close completely and thereby completely turn off $CO_2$ and $H_2O$ exchange in some proportion of the leaf, as Laisk (1983) and Farquhar *et al.* (1987) pointed out, the effect will be similar to a reduction of leaf area, i.e. there will be a proportional reduction in $A$ and $g$ with constant $p_i$ and a depressed $A(p_i)$ and $g(p_i)$ relationship. The question is: is this what happened in Figs 2 and 3? We do not know at present. Terashima *et al.* (1988) showed that the apparent depression by abscisic acid of photosynthetic capacity, as judged by change in $A(p_i)$ relationship, is an artefact of this kind (see Fig. 4). Their experiments were with detached leaves of *Helianthus annuus* and *Vicia faba*, and with attached leaves of *H. annuus*, and involved exogenous application of the hormone. Subsequent, confirmatory experiments by Downton, Loveys & Grant (1988) and Ward & Drake (1988) also used exogenous application. There are numerous reports of

Fig. 4. (*a*) Effect of ABA treatment on relationships between rates of $CO_2$ assimilation and internal partial pressure of $CO_2$, $p_i$, in detached leaves of *Helianthus annuus* L. $CO_2$ and water vapour exchange were measured with an open gas-exchange system, the incident irradiance and leaf temperature were 1000 $\mu$mol photon $m^{-2}$ $s^{-1}$ and 25±1 °C, respectively. The concentration of ABA supplied to the petiole was $10^{-5}$ M (open symbols). Closed symbols denote control leaf (without ABA treatment). The value of stomatal conductance to diffusion of water vapour in mol $m^{-2}$ $s^{-1}$ is given for each datum. (*b*) Photosynthetic oxygen evolution rate as functions of ambient partial pressure of $CO_2$ in a control leaf disc (closed symbol) and on an ABA-treated leaf disc (open symbol) of *Helianthus annuus* L. The ABA-treated leaf disc was obtained from the same ABA-treated detached leaf used in the $CO_2$-exchange measurements depicted in (*a*). For oxygen evolution measurements, the incident irradiance was 1200 $\mu$mol photon $m^{-2}$ $s^{-1}$. These (and other) results suggested that the effects of ABA on the rate of $CO_2$ assimilation depicted in (*a*) are not due to a reduction in the biochemical capacity for photosynthesis, but to heterogeneity in stomatal opening, and can be overcome by imposing high ambient $CO_2$ partial pressure (after Terashima *et al.*, 1988).

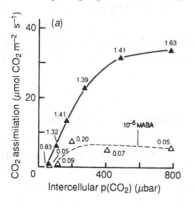
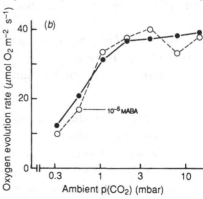

large increases in ABA content in leaves during water stress, but none of whether endogenous ABA causes stomatal heterogeneity as outlined above. It seems likely that the apparent decline in capacity, caused by five days of stress, observed by von Caemmerer & Farquhar (1984), described earlier, was an artefact. The long-term effects of water stress, however, may well involve a real decrease in capacity, particularly if they occur at high light intensity.

The history of the study of the effects of water stress on stomata and photosynthesis is thus an interesting one. Initially all effects were thought to be stomatal in origin, then largely of biochemical nature, but with the lesion being unknown in cases of short-term stress. The short-term effects are now again thought to be, probably, stomatal (but resulting from heterogeneity). For the cases of long-term stress there is greater likelihood of photoinhibition and perhaps of altered proteins.

### How is water stress sensed?

So far we have not considered how either stomata or mesophyll cells sense their change in water status during stress. Traditionally leaf water potential has been used as the measure. The data and discussion of Jones (1973), Kaiser (1987) and Sinclair & Ludlow (1985) emphasised the greater relevance of relative water content, rather than water potential, to the description of dehydration effects. However, there is increasing evidence that soil water status can influence stomatal conductance in ways that are not mediated through any effect on leaf water status (Bates & Hall, 1981; Turner, Schulze & Gollan, 1985; Blackman & Davies, 1985; Gollan, Passioura & Munns, 1986). As the soil dries, it becomes increasingly difficult for roots to penetrate it. We say that the strength of the soil increases. Recently, Masle & Passioura (1987) showed that shoots of seedlings can respond to changes of soil strength *per se*. They invoked hormonal messages from the roots as discussed by Sharp & Davies (Chapter 5).

The diversity of responses to water stress may therefore reflect in part the diversity of signals, from turgor and water content of leaf cells, to root perception of soil water content and strength.

### Transpiration efficiency and carbon isotope discrimination
#### Water-use efficiency

The amount of growth occurring when rainfall is limited depends on the ratio of assimilation rate to transpiration rate. In a leaf the instantaneous transpiration efficiency, $A/E$, is given approximately by

$$\frac{A}{E} = \frac{p_a(1 - p_i/p_a)}{1.6\,v}. \tag{3}$$

The integral of this over the life of the plant, the water-use efficiency $W$ is

$$W = \frac{p_a(1 - p_i/p_a)(1 - \phi_c)}{1.6\,v(1 + \phi_w)}, \tag{4}$$

where $v$ is the mean vapour pressure difference between leaves and air, and $\phi$ represents losses not associated with $CO_2$ uptake through the stomata. For carbon lost by the leaves at night and by roots by respiration we use $\phi_c$, and for water losses by partially closed stomata at night and cuticular losses of water, we use $\phi_w$. Although we here use the term water-use efficiency for the ratio of carbon gained to water lost by a single plant, it is usually reserved in a crop context for the ratio of carbon accumulated to total water used, including soil evaporation.

Early this century, differences in $W$ were demonstrated among species and among cultivars within species (Briggs & Shantz, 1914). These experiments simply involved measurements of plant dry weight and of pot weight. Although simple in concept, they are tedious to apply on the large scale that is required for selection by breeders (see Chapter 11).

### Carbon isotope discrimination

Recently it has been recognised that the carbon isotope composition of small amounts of plant material may be used to assess differences in $p_i/p_a$ and $W$ (Farquhar *et al.*, 1982). There are two stable isotopes of carbon, $^{13}C$ and $^{12}C$, which are in the molar ratio $1:89$ in the atmosphere. During assimilation of atmospheric $CO_2$ the plant fixes a smaller $^{13}C/^{12}C$ ratio than that in the atmosphere. The phenomenon is called carbon isotope discrimination, $\Delta$.

Carbon isotope discrimination by leaves with the $C_3$ pathway of photosynthesis is related to $p_i/p_a$ (Farquhar *et al.*, 1982), through

$$\Delta = a(p_a - p_i)/p_a + b(p_i/p_a) \tag{5}$$

or

$$\Delta = a + (b - a)p_i/p_a, \tag{6}$$

where $a$ $(4.4 \times 10^{-3})$ and $b$ $(30 \times 10^{-3})$ are constants representing the fractionations due to diffusion and carboxylation of $CO_2$. Other potential sources of fractionation and a fuller relationship between $\Delta$ and $p_i/p_a$ has been given by Farquhar & Richards (1984). Substituting the values for $a$ and $b$ into Equation 6 gives a simple expression predicting carbon isotope discrimination

$$\Delta = (4.4 + 25.6\ p_i/p_a) \times 10^{-3}. \tag{7}$$

A positive linear relationship is predicted between $\Delta$ and $p_i/p_a$, and this has been verified for $C_3$ species by Evans *et al.* (1986) and Farquhar *et al.* (1988).

### Isotope discrimination and water-use efficiency

If we rearrange Equation 6 and substitute for $p_i/p_a$ in Equation 4, we obtain a negative dependence of $W$ on $\Delta$:

$$W = \frac{p_a(b - \Delta)(1 - \phi_c)}{1.6(b - a)(v)(1 + \phi_w)}. \tag{8}$$

Negative relationships have now been found for plants in isolated pots in wheat (Farquhar & Richards, 1984), peanut (Hubick, Farquhar & Shorter, 1986; Hubick, Shorter & Farquhar, 1988; Wright, Hubick & Farquhar, 1988), barley (Hubick & Farquhar, 1989; K.T. Hubick, S. von Caemmerer

Fig. 5. Water-use efficiency (carbon basis) *v.* average carbon isotope discrimination in the whole plant, $r = -0.88$. Open symbols represent well-watered plants and closed symbols represent plants that were droughted. ◆, Tifton 8; ▲, Florunner; ✦, VB187 and ✦, Chico are cultivars of peanut (*Arachis hypogaea*). (From Hubick *et al.*, 1986).

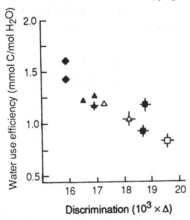

& G.D. Farquhar, unpublished), tomato (Martin & Thorstenson, 1988) and cotton (Hubick & Farquhar, 1987). The sources of variations in $\Delta$ (and $W$) were both environmental and genetic in these experiments. An example is given in Fig. 5.

### Environmental effects on discrimination and water-use efficiency

Any environmental stress that affects leaf conductance and assimilation rate differently, so that the changes in one are not proportional to the changes in the other, will affect $p_i/p_a$ and therefore $\Delta$. If the changes do not affect the terms in the denominator of Equation 8 independently of $p_i/p_a$, then $W$ will also be affected.

The changes in $\Delta$ observed under stress are therefore of interest in the light of the uncertainty presently surrounding gas exchange measurements. Stress imposed on chickpea in small pots caused a large decrease in $\Delta$ (Winter, 1981) whereas long-term (and terminal) stress of wheat caused a reduction of only $1.4 \times 10^{-3}$ (Farquhar & Richards, 1984). From Equation 7 this corresponds to a difference in $p_i/p_a$ of 1.4/25.6, or only 6%. The increase in $W$ was 20%. Of course these represent the integrated differences and include carbon laid down early, before the stress was severe. Hubick *et al.* (1986) reported experiments on different cultivars of peanut and related species. Some of the pots received only enough water for half the rate of loss of the controls. On average there was no effect on either $\Delta$ or $W$. Martin & Thorstenson (1988) kept pots of tomato and two relatives at 100, 50 or 25% of field capacity. Discrimination was decreased by 1.3 and $2.1 \times 10^{-3}$ and $W$ increased by 1 and 17% in the 50 and 25% pots, respectively. Hubick & Farquhar (1989) reported both a decrease in $\Delta$ of $1.3 \times 10^{-3}$ and an increase in $W$ of 17% with water stress of barley genotypes.

In summary, the effects of water stress on $\Delta$ are similar to those discussed earlier for $p_i/p_a$, (i.e. ranging from virtually no response to large decreases). A study involving both gas-exchange and isotope discrimination of plants undergoing stress would be useful, as the changes in $W$ may be complicated by changes in the proportion of carbon lost as respiration.

Over a large range of increasing soil strength Masle & Farquhar (1988) observed a decrease in $\Delta$ of $3.7 \times 10^{-3}$ and a doubling of $W$. The decrease in $\Delta$ was also accompanied by a reduction in the value of $p_i$ measured by gas-exchange.

Salinity has also been shown to decrease discrimination (Guy, Reid & Krouse, 1980; Farquhar *et al.*, 1982; Neales, Fraser & Roksandic, 1983; Downton, Grant & Robinson, 1985; Guy & Reid, 1986; K.T. Hubick & S. von Caemmerer, unpublished). In some cases gas-exchange was measured and $p_i$ found to decrease. Water-use efficiency also increased over the non-salinised control plants.

The effects of abscisic acid (ABA) are also of interest. For example, Bradford, Sharkey & Farquhar (1983) found that its application to tomato leaves during growth reduced $\Delta$ and $p_i$, whereas the mutant, *flacca*, which lacks ABA had greater $\Delta$ and $p_i$ than the wild-type. Other parameters that can affect $p_i/p_a$ and $\Delta$ are irradiance (Francey *et al.*, 1985; Ehleringer *et al.*, 1986), humidity (Winter *et al.*, 1982), altitude (Körner, Farquhar & Roksandic 1987), flooding (Guy & Wample, 1984) and nutrient availability (Bender & Berge, 1979). In an ecological context, differences in $\Delta$ may be useful as an indicator of adaptive advantage for species in water-limited environments. Smith & Osmond (1987) and Ehleringer, Comstock & Cooper (1987) found that $\Delta$ in stem dry matter of desert species which had photosynthetic stems was smaller than $\Delta$ in leaf dry matter, consistent with variation in $p_i/p_a$ (Osmond *et al.*, 1987) and water-use efficiency.

### Genetic variation in discrimination and transpiration efficiency

In the experiments decribed earlier on $\Delta$ and $W$ in wheat, peanut, barley, tomato and cotton, there was a great deal of genetic variation which still fitted the relationship described by Equation 8. The range of $\Delta$ in one species is often about 2 to $4 \times 10^{-3}$ (e.g. Hubick *et al.*, 1986 with peanut, and similarly from surveys of wheat, cotton, barley and cowpea).

Condon, Richards & Farquhar (1987) showed that the ranking of $\Delta$ was similar when 27 cultivars of wheat were grown at two contrasting sites. That is, the genotype × environment interaction was not large. The same conclusions were drawn from studies of peanut (Hubick, Shorter & Farquhar, 1988). An effect of environment on the ranking of $\Delta$ among 18 cowpea genotypes was observed by A.E. Hall, C. Mutters, K.T. Hubick & G.D. Farquhar (unpublished).

Carbon isotope discrimination is a heritable character. Crosses between genotypes of peanut identified as having the combinations of large $W$ and small $\Delta$ or small $W$ and large $\Delta$ resulted in the $F_1$ progeny having $\Delta$ (and $W$) similar to the parent with the small $\Delta$ (Hubick, Shorter & Farquhar, 1988). Isotope discrimination and $W$ were segregating characters and a distribution of both was observed between the parent lines that overlapped with the parents. A strong phenotypic correlation ($r = -0.78$) was found for $W$ and $\Delta$ in the $F_2$ generation. Martin & Thorstenson (1988) made crosses between two tomato species, the cultivated *Lycopersicon esculentum* and the wild *L. pennellii*. In this case $\Delta$ of the $F_1$ generation was intermediate between the parents, but closer to the smaller $\Delta$ (larger $W$) parent. They also found a strong negative correlation between $W$ and $\Delta$ when the parent tomato species and their $F_1$ cross were compared.

## Photoinhibition
### Dealing with high or excessive light
When water is severely limiting, or some other stress decreases photosynthesis, there is the possibility that photoinhibition may exacerbate the problem. Photoinhibition is the damage to photosynthesis by light (see review by Powles, 1984). Photoinhibition can potentially depress growth and reproduction in natural systems, and the yield of crops, by reducing the plant's ability to gain carbon. There is evidence of genetic variation in the ability to deal with high light, and studies have commenced on the relationships with plant distribution and with yield.

### Photochemical efficiency at low irradiance
Light energy that is absorbed by the leaf has several potential fates. The predominant one is to drive the photochemistry that results in regeneration of the NADPH and ATP used in $CO_2$ fixation. When expressed on the basis of absorbed light, that is correcting for reflection and transmission by the leaf, Björkman & Demmig (1987) found no significant variation in the quantum requirement across 37 $C_3$ species from diverse genera. The theoretical quantum requirement for ATP regeneration necessary to fix $CO_2$ is approximately 9 quanta per $CO_2$ molecule. For each $O_2$ molecule evolved, 4 electrons must be transported through both photosystems, which requires $4 \times 2 = 8$ quanta and produces 2 NADPH. In the absence of the Q cycle, $8H^+$ are formed in the thylakoid lumen which could regenerate 2.67 ATP, given that $3H^+$ are required per ATP. However, in the absence of photorespiration, $CO_2$ fixation requires 2 NADPH and 3 ATP, so that the ATP shortfall requires one extra quantum, making 9 in total. If the Q cycle operates, up to $12H^+$ are formed per oxygen evolved, yielding 4 ATP and 2 NADPH per 8 quanta, but other compounds are also being reduced. Plant material is more reduced than carbohydrate, and if the C:H:O:N ratio of leaves, which is approximately 1:1.7:0.61:0.05 (Hubick *et al.*,1986), were to be achieved by photoreactions alone, this requires an extra 0.8 quanta per molecule of $CO_2$, giving 8.8 in total (Farquhar, 1988). Measurements of gas exchange by leaves have demonstrated that, for wavelengths around 600 nm, about 9 quanta are required to fix each molecule of $CO_2$ (see Evans, 1987). This is only possible if light is equally distributed between the two photosystems, because they operate in series and are spatially separated (Anderson, 1986). It also means that there are negligible losses by other processes such as non-radiative decay or fluorescence.

At low irradiances, photosynthesis uses virtually 100% of the quanta, but in full sunlight, about 2000 $\mu$mol quanta m$^{-2}$ s$^{-1}$, more quanta are available than can be used in photochemistry. Maximum rates of photosynthesis by *Populus* or *Spinacia* leaves of 15 and 70 $\mu$mol $O_2$ m$^{-2}$ s$^{-1}$, respectively, would require only 15 $\times$ 9 = 135 to 630 $\mu$mol quanta m$^{-2}$ s$^{-1}$, or 10–40%. Leaves, therefore, need to be able to dissipate 60–90% of the quanta at high irradiance in an orderly manner such as non-radiative decay if they are to avoid the potentially damaging formation of oxygen radicals from reduced ferredoxin (Asada & Takahashi, 1987). When plants are under a stress that restricts $CO_2$ assimilation, excessive light will be reached at even lower irradiances.

### Importance of leaf movements

Solar tracking by leaves through the day can increase the daily receipt of light. However, in times of stress excessive light can be avoided by leaf movements. The apex of the leaf can be oriented parallel to the sun (legumes: Ludlow & Björkman, 1984), or the leaf can roll up (monocots) or wilt (dicots), to reduce greatly the intercepted irradiance. Some species can also re-orient chloroplasts within the cells to reduce their light interception (e.g. *Oxalis*). Not all species have these avoidance mechanisms. The ones without them need to rely instead on biochemical mechanisms to dissipate excess energy. In fact, these mechanisms probably also function in species with avoidance tactics.

### Non-radiative energy dissipation

The second major pathway for absorbed light, after photochemistry, is non-radiative decay. In Powles' (1984) review of photoinhibition, he emphasised that a better understanding of the quantitative contribution of non-radiative dissipation of light was needed. This has been helped recently by the development of techniques which enable leaf fluorescence to be analysed (e.g. Schreiber, 1986). While fluorescence itself accounts for less than 1% of absorbed light, it reflects the state of the photosystem II reaction centres. Consequently, it has been a useful indicator of photoinhibition because the ratio of variable to maximal fluorescence $F_v/F_m$ correlates with the quantum yield of $O_2$ evolution (e.g. Björkman, 1987). Fluorescence is reduced by several processes, photochemistry or Q quenching, and energy-dependent quenching associated with either the $\Delta$pH across the thylakoid membrane or non-radiative dissipation of energy.

In the short term (minutes), energy-dependent quenching reflects the buildup of the trans-thylakoid proton gradient. When the $\Delta$pH is high, there is an increase in the orderly dissipation of energy that protects photosystem II from photoinhibition. This was demonstrated by partially uncoupling thylakoids to reduce the pH gradient without inhibiting $CO_2$-dependent oxygen evolution, with the result that photoinhibition occurred (Krause & Behrend, 1986). While Krause & Laasch (1987) and Demmig & Winter (1988a) argue that a loss of quantum yield is primarily related to energy-dependent quenching and not the reduction state of the reaction centre, Weis & Berry (1987) argue that it is the quantum yield of open reaction centres that is reduced by energy-dependent quenching. In any event, the mechanism by which energy is dissipated in the short term as a result of the $\Delta$pH gradient is obscure.

*Xanthophyll cycle*

Over longer times (30 min), energy-dependent quenching may involve a carotenoid cycle (Demmig & Björkman, 1987). Carotenoids are the second most prominent pigment after chlorophylls, contributing to both light absorption and energy dissipation. The photosystem reaction centres contain carotene (14–20 per 100 Chl), whereas the light-harvesting chlorophyll *a/b*–protein complexes contain xanthophylls (*c.* 25 per 100 Chl) (Siefermann-Harms, 1985). The presence of carotenes is essential for the retention of chlorophyll as mutants lacking them die in the light (Anderson & Robertson, 1960). If electron transfer from reaction centres is blocked, charge recombination results in the formation of the triplet excited state of chlorophyll which can be safely dissipated as heat by reactions with carotenoids (Asada & Takahashi, 1987). However, it is the xanthophylls in the light-harvesting complex which undergo a reversible, light-induced interconversion called the xanthophyll cycle (Hager, 1980). Violaxanthin is converted to zeaxanthin by a de-epoxidase which is a water-soluble enzyme of MW 54 kDa, active at pH 5 (Hager & Perz, 1970). The back reaction is catalysed by an oxygenase with a pH optimum of 7.6, consuming NADPH and oxygen. Zeaxanthin presumably differs from violaxanthin in that it readily loses light energy as heat, although this has yet to be demonstrated.

This xanthophyll cycle was proposed as a mechanism for dissipating excess light energy because the amount of zeaxanthin present correlated with a measure of non-radiative dissipation of energy by photosystem II in *Populus balsamifera* and *Hedera helix* (Demmig et al., 1987). As light intensity increases above moderate levels, the rate of electron transport cannot keep pace with the rate of receipt of quanta. The xanthophyll cycle provides a mechanism for the orderly dissipation of energy in excess of that required by photosynthesis through the involvement of the membrane pH gradient. When the rate of use of protons by the ATPase is insufficient, the pH gradient increases, activating the de-epoxidase. In turn, the conversion of violaxanthin to zeaxanthin leads to a greater non-radiative energy dissipation in the membranes and reduced proton deposition in the thylakoid lumen. The cycle is self-regulating, because as electron transport slows due to fewer quanta reaching the reaction centres, the pH gradient declines and zeaxanthin is converted back to violaxanthin. By dissipating energy as heat from the pigment bed, the potentially damaging formation of oxygen radicals from reduced ferredoxin is avoided.

Under conditions of excess light, the carotenoid cycle gradually replaces the other more rapid energy-dependent quenching process (Demmig & Winter, 1988b) and damage may only be evident if the capacity of the cycle is exceeded. Whereas depression of quantum yield and 77 K variable fluorescence from photosystem II was considered as evidence of photoinhibition (Powles & Björkman, 1982b; Demmig & Björkman, 1987), this could reflect in part the protective carotenoid cycle which relaxes slowly. Having a mechanism for protection against excess light should greatly facilitate selection where photoinhibition is a problem.

The non-radiative dissipation of light converts the energy into heat. There is also a considerable amount of heat produced during photochemistry. Quanta of blue (400 nm) or red (700 nm) light contain 0.30 or 0.17 MJ $mol^{-1}$, while the formation of NADPH and ATP requires 0.22 and 0.33 MJ $mol^{-1}$, respectively. Thus the 9 mol of quanta needed to fix 1 mol of $CO_2$ contain 1.54–2.69 MJ and regenerate 2.25 NADPH and 3 ATP, $(2.25 \times 0.22 + 3 \times 0.03 = 0.59$ MJ$)$, or 38–22% efficiency. The remaining 62–78% is lost as heat. A further 10% is lost in the Calvin cycle in the absence of photorespiration. This loss to heat is distinct from the 60–90% of quanta at high irradiance that needs to be dissipated as heat to avoid damage, and is the consequence of making photochemistry irreversible.

*Photoinhibition in relation to yield and species distribution*
There is as yet little evidence that photoinhibition reduces crop yield (Ludlow, 1987). Perhaps the most compelling evidence comes from photoinhibition induced by chilling temperature (Oquist, Greer & Ogren, 1987). Shadecloth used over the winter period to reduce photoinhibition increased the yield of tea (Aoki, 1986). For both maize (Long, East & Baker, 1983; Farage & Long, 1987) and kiwifruit, *Actinidia deliciosa* (Greer, 1988), photoinhibition could be induced by chilling temperatures. However, it was not clear to what extent this would be associated with yield reduction, nor whether photoinhibition would be the primary cause of the reduction. Ludlow & Powles (1988) tried to induce photoinhibition by water stress; however, no effect of photoinhibition could be observed due to

the severity of the water stress. Similarly, drought did not result in photosystem II photoinhibition in cotton, although there was some reduction in whole-chain electron transport (Genty, Briantais & Da Silva, 1987). It is interesting to calculate the expected reduction in electron transport rate that results from stomatal closure in a $C_3$ leaf. Reducing leaf conductance by 40% reduces the rate of $CO_2$ assimilation by 12% and intercellular $CO_2$ partial pressure by 18%, but scarcely alters the necessary rate of electron transport (2%) because of increased photorespiration. Increased photorespiration was observed in water-stressed sunflower (Lawlor & Fock, 1975). Photoinhibition is therefore unlikely to occur before considerable water stress is encountered.

The distribution of species is strongly influenced by the light environment and many characteristics of leaves from sun and shade plants have been examined (Boardman, 1977). Leaves from shade plants devote more protein to complexing pigments at the expense of soluble protein. They generally have low maximum rates of photosynthesis and low respiration rates which enable them to reach a positive carbon balance at lower irradiances. Sun and shade clones have been compared in several species (*Solidago*: Björkman & Holmgren, 1963; *Solanum*: Gauhl, 1976). However, in the presence of adequate nitrogen nutrition, no differences between clones of *Solanum* were apparent as all were able to acclimate to high irradiance (Osmond, 1983; Ferrar & Osmond, 1986). Nitrogen stress clearly exacerbates the susceptibility to photoinhibition such that under natural conditions, plants would be confined to shaded sites if soil fertility was low. As always, there are exceptions and even with good mineral nutrition, some plants suffer chronic photoinhibition when grown in high irradiance (Anderson & Osmond, 1987).

### References

Anderson, I.C. & Robertson, D.S. (1960). Role of carotenoids in protecting chlorophyll from photodestruction. *Plant Physiology*, **35**, 531–4.

Anderson, J.M. (1986). Photoregulation of the composition, function, and structure of thylakoid membranes. *Annual Review of Plant Physiology*, **37**, 93–136.

Anderson, J.M. & Osmond, C.B. (1987). Shade–sun responses: compromises between acclimation and photoinhibition. In *Photoinhibition*, ed. D.J. Kyle, C.B. Osmond and C.J. Arntzen, pp. 1–38. Amsterdam: Elsevier.

Aoki, S. (1986). Interaction of light and low temperature in depression of photosynthesis in tea leaves. *Japanese Journal of Crop Science*, **55**, 496–503.

Asada, K. & Takahashi, M. (1987). Production and scavenging of active oxygen in photosynthesis. In *Photoinhibition*, ed. D.J. Kyle, C.B. Osmond and C.J. Arntzen, pp. 227–87. Amsterdam: Elsevier.

Bates, L.M. & Hall, A.E. (1981). Stomatal closure with soil water depletion not associated with changes in bulk leaf water status. *Oecologia*, **50**, 62–5.

Bender, M.M. & Berge, A.J. (1979). Influence of N and K fertilization and growth temperature on $^{13}C$: $^{12}C$ ratios of timothy (*Phleum pratense* L.). *Oecologia*, **44**, 117–18.

Ben-Zioni, A., Itai, C. & Vaadia, Y. (1967). Water and salt stresses, kinetin and protein synthesis in tobacco leaves. *Plant Physiology*, **42**, 361–5.

Björkman, O. (1987). Low-temperature chlorophyll fluorescence in leaves and its relationship to photon yield of photosynthesis in photoinhibition. In *Photoinhibition*, ed. D.J. Kyle, C.B. Osmond and C.J. Arntzen, pp. 123–44. Amsterdam: Elsevier.

Björkman, O. & Demmig, B. (1987). Photon yield of oxygen evolution and chlorophyll fluorescence characteristics at 77 K among vascular plants of diverse origins. *Planta*, **170**, 489–504.

Björkman, O. & Holmgren, P. (1963). Adaptability of the photosynthetic apparatus to light intensity in ecotypes from exposed and shaded habitats. *Physiologia Plantarum*, **16**, 889–914.

Blackman, P.G. & Davies, W.J. (1985). Root to shoot communication in maize plants of the effects of soil drying. *Journal of Experimental Botany*, **36**, 39–48.

Boardman, N.K. (1977). Comparative photosynthesis of sun and shade plants. *Annual Review of Plant Physiology*, **28**, 355–77.

Boyer, J.S. (1970). Differing sensitivity of photosynthesis to low leaf water potential in corn and soybean. *Plant Physiology*, **46**, 236–9.

Bradford, K.J., Sharkey, T.D. & Farquhar, G.D. (1983). Gas exchange, stomatal behaviour, and $\delta^{13}C$ values of the *flacca* tomato mutant in relation to abscisic acid. *Plant Physiology*, **72**, 245–50.

Briggs, L.J. & Shantz, H.L. (1914). Relative water requirement of plants. *Journal of Agricultural Research*, **3**, 1–64.

Brix, H. (1962). The effect of water stress on the rates of photosynthesis and respiration in tomato plants and loblolly pine seedlings. *Physiologia Plantarum*, **15**, 10–20.

Caemmerer, S. von & Farquhar, G.D. (1984). Effects of partial defoliation, changes of irradiance during growth, short-term water stress and growth at enhanced p(CO$_2$) on the photosynthetic capacity of leaves of *Phaseolus vulgaris* L. *Planta*, **160**, 320–9.

Condon, A.G., Richards, R.A. & Farquhar, G.D. (1987). Carbon isotope discrimination is positively correlated with grain yield and dry matter production in field-grown wheat. *Crop Science*, **27**, 996–1001.

Demmig, B. & Björkman, O. (1987). Comparison of the effect of excessive light on chlorophyll fluorescence (77 K) and photon yield of O$_2$ evolution in leaves of higher plants. *Planta*, **171**, 171–84.

Demmig, B. & Winter, K. (1988a). Light response of CO$_2$ assimilation, reduction state of Q and radiationless energy dissipation in intact leaves. In *Ecology of Photosynthesis in Sun and Shade*, ed. J.R. Evans, S. von Caemmerer and W.W. Adams III, pp. 151–62. Melbourne: CSIRO.

Demmig, B. & Winter, K. (1988b). Characterisation of three components of non-photochemical fluorescence quenching and their response to photoinhibiton. In *Ecology of Photosynthesis in Sun and Shade*, ed. J.R. Evans, S. von Caemmerer and W.W. Adams III, pp. 163–78. Melbourne: CSIRO.

Demmig, B., Winter, K., Kruger, A. & Czygan, F.C. (1987). Photoinhibition and zeaxanthin formation in intact leaves. A possible role of the xanthophyll cycle in the dissipation of excess light energy. *Plant Physiology*, **84**, 218–24.

Downton, W.J.S., Grant, J.R. & Robinson, S.P. (1985). Photosynthetic and stomatal response of spinach leaves to salt stress. *Plant Physiology*, **77**, 85–8.

Downton, W.J.S., Loveys, B.R. & Grant, W.J.R. (1988). Stomatal closure fully accounts for the inhibition of photosynthesis by abscisic acid. *New Phytologist*, **108**, 263–6.

Ehleringer, J.R., Comstock, J.P. & Cooper, T.A. (1987). Leaf twig carbon isotope ratio differences in photosynthetic-twig desert shrubs. *Oecologia*, **71**, 318–20.

Ehleringer, J.R., Field, C.B., Lin, Z.F. & Kuo, C.Y. (1986). Leaf carbon isotope and mineral composition in subtropical plants along an irradiance cline. *Oecologia*, **70**, 520–6.

Evans, J.R. (1987). The dependence of quantum yield on wavelength and growth irradiance. *Australian Journal of Plant Physiology*, **14**, 69–79.

Evans, J.R., Sharkey, T.D., Berry, J.A. & Farquhar, G.D. (1986). Carbon isotope discrimination measured concurrently with gas exchange to investigate $CO_2$ diffusion in leaves of higher plants. *Australian Journal of Plant Physiology* **13**, 281–92.

Farage, P.K. & Long, S.P. (1987). Damage to maize photosynthesis in the field during periods when chilling is combined with high photon fluxes. In *Progress in Photosynthesis Research*, ed. J. Biggins, pp. IV.2.139–42. Dordrecht: Martinus Nijhoff.

Farquhar, G.D. (1988). Models relating subcellular effects of temperature to whole plant responses. In *Plants and Temperature*, ed. S.P. Long and F.I. Woodward. *Society for Experimental Biology Symposium 42* (in press).

Farquhar, G.D., Ball, M.C., von Caemmerer, S. & Roksandic, Z. (1982). Effect of salinity and humidity of $\delta^{13}C$ value of halophytes – evidence for diffusional isotope fractionation determined by the ratio of intercellular/atmospheric partial pressure of $CO_2$ under different environmental conditions. *Oecologia*, **52**, 121–4.

Farquhar, G.D., Hubick, K.T., Condon, A.G. & Richards, R.A. (1988). Carbon isotope fractionation and plant water-use efficiency. In *Applications of Stable Isotope Ratios to Ecological Research*, ed. P.W. Rundel, J.R. Ehleringer & K.A. Nagy, pp.21–40. New York: Springer-Verlag.

Farquhar, G.D., Hubick, K.T., Terashima, I., Condon, A.G. & Richards, R.A. (1987). Genetic variation in the relationship between photosynthetic $CO_2$ assimilation rate and stomatal conductance to water loss. In *Progress in Photosynthesis Research*, Vol. IV, ed. J. Biggins, pp. 5: 209–12. Dordrecht: Martinus Nijhoff.

Farquhar, G.D. & Richards, R.A. (1984). Isotopic composition of plant carbon correlates with water-use efficiency of wheat genotypes. *Australian Journal of Plant Physiology*, **11**, 539–52.

Ferrar, P.J. & Osmond, C.B. (1986). Nitrogen supply as a factor influencing photoinhibition and photosynthetic acclimation after transfer of shade-grown *Solanum dulcamara* to bright light. *Planta*, **168**, 563–70.

Francey, R.J., Gifford, R.M., Sharkey, T.D., Weir, B. (1985). Physiological influences on carbon isotope discrimination in huon pine (*Lagarostrobes franklinii*). *Oecologia*, **66**, 211–18.

Gauhl, E. (1976). Photosynthetic response to varying light intensity in ecotypes of *Solanum dulcamara* L. from shaded and exposed habitats. *Oecologia*, **22**, 275–86.

Genty, B., Briantais, J.M. & Da Silva, J.B.V. (1987). Effects of drought on primary photosynthetic processes of cotton leaves. *Plant Physiology*, **83**, 360–4.

Gollan, T., Passioura, J.B. & Munns, R. (1986). Soil water status affects the stomatal conductance of fully turgid wheat and sunflower leaves. *Australian Journal of Plant Physiology*, 13, 459–64.

Greer, D.H. (1988). Effect of temperature on photoinhibition and recovery in *Actinidia deliciosa*. In *Ecology of Photosynthesis in Sun and Shade*, ed. J.R. Evans, S. von Caemmerer and W.W. Adams III, pp. 195–206. Melbourne: CSIRO.

Guy, R.D., Reid, D.M. & Krouse, H.R. (1980). Shifts in carbon isotope ratios of two $C_3$ halotypes under natural and artificial conditions. *Oecologia*, 44, 241–7.

Guy, R.D. & Reid, D.M. (1986). Photosynthesis and the influence of $CO_2$-enrichment. *Plant, Cell and Environment*, 9, 65–72.

Guy, R.D. & Wample, R.L. (1984). Stable carbon isotope ratios of flooded and non- flooded sunflowers (*Helianthus annuus*). *Canadian Journal of Botany*, 62, 1770–4.

Hager, A. (1980). The reversible, light-induced conversion of xanthophylls in the chloroplast. In *Pigments in Plants*, ed. F.C. Czygan, pp. 57–79. Stuttgart: Gustav Fischer-Verlag.

Hager, A & Perz, H. (1970). Veranderung der lichtabsorption eines carotenoids im enzym (de-epoxidase)-substrat (violaxanthin)-komplex. *Planta*, 93, 314–22.

Hsiao, T.C. (1973). Plant responses to water stress. *Annual Review of Plant Physiology*, 24, 519–70.

Hubick, K.T. & Farquhar, G.D. (1987). Carbon isotope discrimination – selecting for water use efficiency. *Australian Cotton Grower* 8, 66–8.

Hubick, K.T. & Farquhar, G.D. (1989). Genetic variation of transpiration efficiency among barley genotypes is negatively correlated with carbon isotope discrimination. *Plant, Cell and Environment* (in press).

Hubick, K.T., Farquhar, G.D. & Shorter,R. (1986). Correlation between water-use efficiency and carbon isotope discrimination in diverse peanut (*Arachis*) germplasm. *Australian Journal of Plant Physiology*, 13, 803–16.

Hubick, K.T., Shorter, R. & Farquhar, G.D. (1988). Heritability and genotype × environment interactions of carbon isotope discrimination and transpiration efficiency in peanut. *Australian Journal of Plant Physiology*, 15 (in press).

Ishihara, K., Nishihara, T. & Ogura, T. (1971). The relationship between environmental factors and behaviour of stomata in the rice plant. I. On the measurement of the stomatal aperture. *Proceedings of the Japanese Society for Crop Science*, 40, 491–6.

Itai, C. & Vaadia, Y. (1965). Kinetin-like activity in root exudate of water-stressed sunflower plants. *Physiologia Plantarum*, 18, 941–4.

Itai, C. & Vaadia, Y. (1971). Cytokinin activity in water-stressed shoots. *Plant Physiology*, 47, 87–90.

Jarvis, P.G. & McNaughton, K.G. (1985). Stomatal control of transpiration: Scaling up from leaf to region. *Advances in Ecological Research*, 15, 1–49.

Jones, H.G. (1973). Photosynthesis by thin leaf slices in solution. II. Osmotic stress and its effects on photosynthesis. *Australian Journal of Biological Sciences*, 26, 25–33.

Kaiser, W.M. (1982). Correlation between changes in photosynthetic activity and changes in total protoplast volume in leaf tissue from hygro-, meso-, and xerophytes under osmotic stress. *Planta*, 154, 538–45.

Kaiser, W.M. (1987). Effects of water deficit on photosynthetic capacity. *Physiologia Plantarum*, **71**, 142–9.

Körner, C., Farquhar, G.D. & Roksandic, Z. (1987). A global survey of carbon isotope discrimination in plants from high altitude. *Oecologia*, **72**, 1–10.

Krause, G.H. & Behrend, U. (1986). $\Delta$ pH-dependent chlorophyll fluorescence quenching indicating a mechanism of protection against photoinhibition of chloroplasts. *FEBS Letters*, **200**, 298–302.

Krause, G.H. & Laasch, H. (1987). Energy-dependent chlorophyll fluorescence quenching in chloroplasts correlated with quantum yield of photosynthesis. *Zeitschrift für Naturforschung*, **42**, 581–4.

Laisk, A. (1983). Calculation of photosynthetic parameters considering the statistical distribution of stomatal apertures. *Journal of Experimental Botany*, **34**, 1627–35.

Laisk, A., Oja, V. & Kull, K. (1980). Statistical distribution of stomatal apertures of *Vicia faba* and *Hordeum vulgare* and Spannungsphase of stomatal opening. *Journal of Experimental Botany*, **31**, 49–58.

Lawlor, D.W. & Fock, H. (1975). Photosynthesis, photorespiratory $CO_2$ evolution of water stressed sunflower leaves. *Planta*, **126**, 247–58.

Lawlor, D.W. & Fock, H. (1978). Photosynthesis, respiration and carbon assimilation in water-stressed maize at two oxygen concentrations. *Journal of Experimental Botany*, **29**, 579–93.

Long, S.P., East, T.M. & Baker, N.R. (1983). Chilling damage to photosynthesis in young *Zea mays*. 1. Effects of light and temperature variation on photosynthetic $CO_2$ assimilation. *Journal of Experimental Botany*, **34**, 177–88.

Ludlow, M.M. (1987). Light stress at high temperature. In *Photoinhibition*, ed. D.G. Kyle, C.B. Osmond and C.J. Arntzen, pp. 89–109. Amsterdam: Elsevier.

Ludlow, M.M. & Björkman, O. (1984). Paraheliotropic leaf movement in *Siratro* as a protective mechanism against drought-induced damage to primary photosynthetic rections: damage by excessive light and heat. *Planta*, **161**, 505–18.

Ludlow, M.M. & Powles, S.B. (1988). Effects of photoinhibition induced by water stress of growth and yield of grain sorghum. In *Ecology of Photosynthesis in Sun and Shade*, ed. J.R. Evans, S. von Caemmerer and W.W. Adams III, pp. 179–94. Melbourne: CSIRO.

Martin, B.J. & Thorstenson, Y.R. (1988). Stable carbon isotope composition ($\delta^{13}C$), water use efficiency and biomass productivity of *Lycopersicon esculentum*, *Lycopersicon pennellii* and the $F_1$ hybrid. *Plant Physiology* **88**, 218–23.

Masle, J. & Farquhar, G.D. (1988). Effects of soil strength on the relation of water use efficiency and growth to carbon isotope discrimination in wheat seedlings. *Plant Physiology*, **6**, 147–55.

Masle, J. & Passioura, J.B. (1987). Effects of soil strength on the growth of wheat seedlings. *Australian Journal of Plant Physiology*, **141**, 643–56.

Neales, T.F., Fraser, M.S. & Roksandic, Z. (1983). Carbon isotope composition of the halophyte *Disphyma clavellatum* (Haw.) Chinnock (Aizoaceae), as affected by salinity. *Australian Journal of Plant Physiology*, **10**, 437–44.

Oquist, G., Greer, D.H. & Ogren, E. (1987). Light stress at low temperature. In *Photoinhibition*, ed. D.J. Kyle, C.B. Osmond and C.J. Arntzen, pp. 67–87. Amsterdam: Elsevier.

68    G.D. Farquhar et al.

Osmond, C.B. (1983). Interactions between irradiance, nitrogen nutrition and water stress in the sun-shade responses of *Solanum dulcamara*. *Oecologia*, **57**, 316–21.

Osmond, C.B., Smith, S.D., Ben, G.-Y. & Sharkey, T.D. (1987). Stem photosynthesis in a desert ephemeral, *Eriogonum inflatum*: characterization of leaf and stem $CO_2$ fixation and $H_2O$ vapour exchange under controlled conditions. *Oecologia*, **72**, 542–49.

Osmond, C.B., Winter, K. & Powles, S.B. (1979). Adaptive significance of $CO_2$ cycling in leaves of plants with different photosynthetic pathways. In *Stress Physiology*, ed. N.C. Turner and P.J. Kramer. New York: Academic Press.

Potter, J.R. & Boyer, J.S. (1973). Chloroplast response to low leaf water potentials. II. Role of osmotic potential. *Plant Physiology*, **51**, 993–7.

Powles, S.B. (1984). Photoinhibition of photosynthesis induced by visible light. *Annual Review of Plant Physiology*, **35**, 15–44.

Powles, S.B. & Björkman, O. (1982a). High light and water stress effects on photosynthesis in *Nerium oleander*. II. Inhibition of photosynthetic reaction under water stress: interaction with light level. *Carnegie Institute of Washington Year Book*, **81**, 76–7.

Powles, S.B. & Björkman, O. (1982b). Photoinhibition of photosynthesis: effect on chlorophyll fluorescence at 77 K in intact leaves and chloroplast membranes of *Nerium oleander*. *Planta*, **156**, 97–107.

Schreiber, U. (1986). Detection of rapid induction kinetics with a new type of high frequency modulated chlorophyll fluorometer. *Photosynthesis Research*, **9**, 261–72.

Sharkey, T.D. & Badger, M.R. (1982). Effects of water stress on photosynthetic electron transport, photophosphorylation and metabolite levels of *Xanthium strumarium* mesophyll cells. *Planta*, **156**, 199–206.

Siefermann-Harms, D. (1985). Carotenoids in photosynthesis. I. Location in photosynthetic membranes and light-harvesting function. *Biochimica Biophysica Acta*, **811**, 325–55.

Sinclair, T.R. & Ludlow, M.M. (1985). Who taught plants thermodynamics? The unfulfilled potential of plant water potential. *Australian Journal of Plant Physiology*, **12**, 213–17.

Smith, B.N. & Epstein, S. (1971). Two categories of $^{13}C/^{12}C$ ratios for higher plants. *Plant Physiology*, **47**, 380–4.

Smith, S.D. & Osmond, C.B. (1987). Stem photosynthesis in a desert ephemeral, *Eriogonum inflatum*. *Oecologia*, **72**, 533–41.

Takeda, T., Sugimoto, H. & Agata, W. (1978). Water and crop production. I. The relationship between photosynthesis and transpiration in corn leaf. *Japanese Journal of Crop Science*, **47**, 82–9.

Terashima, I., Wong, S.D., Osmond, C.B. & Farquhar, G.D. (1988). Characterisation of non-uniform photosynthesis induced by abscisic acid in leaves having different mesophyll anatomies. *Plant Cell Physiology*, **29** 385–94.

Turner, N.C., Schulze, B.-D. & Gollan, T. (1985). The response of stomata and leaf gas exchange to vapour pressure deficits and soil water content. II. In the mesophytic herbaceous species *Helianthus annuus*. *Oecologia*, **65**, 348–55.

van den Driessche, R., Connor, D.J. & Tunstall, B.R. (1971). Photosynthetic response of brigalow to irradiance, temperature and water potential. *Photosynthetica*, **5**, 210–17.

Ward, D.A. & Drake, B.G. (1988). Osmotic stress temporarily reverses the inhibitions of photosynthesis and stomatal conductance by abscisic acid – evidence that abscisic acid induces a localized closure of stomata in intact, detached leaves. *Journal of Experimental Botany*, **39**, 147–55.

Weis, E. & Berry, J.A. (1987). Quantum efficiency of photosystem II in relation to 'energy'-dependent quenching of chlorophyll fluorescence. *Biochimica Biophysica Acta*, **894**, 198–208.

Winter, K. (1981). $CO_2$ and water vapour exchange, malate content and $\delta^{13}C$ values in *Cicer arietinum* growth under two water regimes. *Zeitschrift für Pflanzenphysiologie*, **101**, 421–30.

Winter, K., Holtum, J.A.M., Edwards, G.E. & O'Leary, M.H. (1982). Effect of low relative humidity on $\delta^{13}C$ values in two $C_3$ grasses. *Journal of Experimental Botany*, **33**, 88–91.

Wong, S.C., Cowan, I.R. & Farquhar, G.D. (1985). Leaf conductance in relation to rate of $CO_2$ assimilation. III. Influence of water stress and photoinhibition. *Plant Physiology*, **78**, 830–4.

Wright, G.C., Hubick, K.T. & Farquhar, G.D. (1988). Discrimination in carbon isotopes of leaves correlates with water-use efficiency of field-grown peanut cultivars. *Australian Journal of Plant Physiology*, **15** (in press).

R.E. SHARP AND W.J. DAVIES

# 5 Regulation of growth and development of plants growing with a restricted supply of water

## Introduction

On a global basis, drought limits plant growth and crop productivity more than any other single environmental factor (Boyer, 1982). Even in Britain, rain-free periods are frequent enough for irrigation to lead to yield advantages for many agricultural and horticultural crops. Water supply is restricted in many parts of the world and productivity in these environments can only be increased by the development of crops that are well adapted to dry conditions. It is clear that the potential for biotechnological improvement of crop performance cannot be realised until we have identified genes and gene products which are responsible for the desired characteristics of drought tolerance. This in turn cannot occur without a thorough understanding of the biophysical, biochemical and physiological perturbations that are induced by a restricted water supply.

Although plant growth rates are generally reduced when soil water supply is limited, shoot growth is often more inhibited than root growth and in some cases the absolute root biomass of plants in drying soil may increase relative to that of well-watered controls (Sharp & Davies, 1979; Malik, Dhankar & Turner, 1979). It is also commonly observed (e.g. Sharp & Davies, 1985) that the roots of unwatered plants grow deeper into the soil than roots of plants that are watered regularly. Clearly, increases in the density and depth of rooting can help sustain a high rate of water extraction in drying soil (Sharp & Davies, 1985) and may promote substantial improvement in yield in dry years (Jordan, Dugas & Shouse, 1983). There is still considerable uncertainty over the underlying causes of drought-induced changes in plant development and this chapter describes how the growth of shoots and roots of different plants might be regulated as a function of changes in water availability.

### Relative sensitivity to water deficit of the expansive growth of different plant parts

Expansive growth is considered to be one of the most sensitive of plant processes to the development of water deficits (Bradford & Hsiao, 1982). It has generally been assumed that when soil water supply is limited shoot growth is more inhibited than root growth because of exposure to the dehydrating effects of the atmosphere. A recent study by Westgate & Boyer (1985a) showed, however, that the growth of maize nodal roots is intrinsically less sensitive than that of the aerial parts of the plant to low water potentials in the growing region (Fig. 1). The water potential at which growth ceased was − 0.50, − 0.75, and − 1.00 MPa in stems, silks and leaves, respectively, while root growth continued to a water potential of − 1.4 MPa.

Cell enlargement occurs when a demand for water is created by relaxation of the cell walls under the influence of turgor pressure and wall-loosening factors. Water enters the cell down a water potential gradient, extending the cell walls (Lockhart, 1965; Boyer, 1985; Tomos, 1985).

Fig. 1. Elongation rate of stem internode 12 (▲), silks (△), leaf 8 (●), and nodal roots (○) of maize at various water potentials. Elongation rates are the average per hour for 24 h of growth in a controlled environment chamber. Water potentials were measured in the growing region of each organ in the same plants. Samples were taken immediately after the growth period when the plants had been in the dark for the last 10 h. Each point is from a single plant. Modified from Westgate & Boyer (1985a).

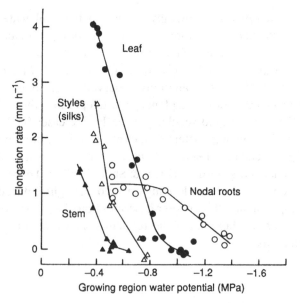

Solutes must also enter to provide the metabolites for cell wall synthesis and to maintain the osmotic forces necessary to drive enlargement. A restricted supply of water might induce changes in any of these processes.

Traditionally, the effects of low water potential on expansive growth have been attributed to turgor losses. A linear relationship between growth rate and turgor can be demonstrated for many plant parts (see e.g. Boyer (1968) for growing leaves). In many reports where leaf growth rate apparently declines with declining turgor, however, turgor was actually measured on pieces of the non-growing lamina. Interestingly, when measurements are made in the regions of leaves and stems where cell enlargement is taking place, turgor often shows little or no decrease even though enlargement is inhibited substantially as a result of soil drying (Meyer & Boyer, 1972; 1981; Matsuda & Riazi, 1981; Michelena & Boyer, 1982; Westgate & Boyer, 1985*a*). This is because, in these cases, solutes accumulate in the growing cells as their water potential falls. This process has become known as osmotic adjustment (Turner & Jones, 1980; Morgan, 1984), and it was originally suggested that because of the resulting maintenance of turgor, such adjustment allowed plants to maintain growth at low water potentials. Clearly, however, in leaves and stems at least, solute accumulation does not fully compensate for the effects of limited water supply on cell enlargement. Indeed, in these organs, solutes apparently 'pile up' *because* of the inhibition of growth (e.g. Meyer & Boyer, 1981; Van Volkenburgh & Boyer, 1985). In contrast, inhibition of silk growth in maize at low water potentials (Fig. 1) is apparently quite tightly linked to losses in turgor due to lack of osmotic adjustment. Failure of reproduction in maize under dry conditions is often attributed to the inhibition of silk elongation (Herrero & Johnson, 1981). The lack of osmotic adjustment in this organ may be related to low availability of photosynthate at this stage of development (Westgate & Boyer, 1985*b*).

Westgate & Boyer (1985*a*) suggest that the complete restriction of leaf growth of maize at low water potential, despite turgor maintenance, occurs because the plant is unable to sustain a water potential gradient between the growing regions of the leaf and the xylem and, therefore, cannot sustain water supply to the growing cells. The water potential gradient into the growing region and the consequent supply of water is sustained in the nodal roots on the same plant, however, and in this case the maintenance of turgor by osmotic adjustment allows growth to continue. This difference occurs because the xylem water potential decreases more in the leaf than in the root, indicating the development of a high flow resistance in the vascular system. The role of osmotic adjustment in the maintenance of root elongation at low water potentials is discussed in more detail below.

It seems that growth inhibition at low water potentials is also attributable to alterations in the cell wall yielding process. Data on the effects of soil drying on wall properties are few and far between, but Matthews, Van Volkenburgh & Boyer (1984) and Van Volkenburgh & Boyer (1985) suggest that drought-induced decreases in wall extensibility do in part limit leaf growth in sunflower and maize. Recently, changes in cell wall proteins induced by low water potentials have been reported in soybean stems (Bozarth, Mullet & Boyer, 1987), although the relationship of these changes to growth inhibition is not clear.

We have reported above that osmotic adjustment at low water potential in the growing regions of leaves and stems seems to occur as a result of a reduction in growth. It should be noted, however, that osmotic adjustment is not inevitable when growth slows down (e.g. Meyer & Boyer, 1972) and it has been argued, therefore, that the process does serve to maintain slow growth of these organs at low water potentials (but see Steponkus, Shahan & Cutler (1982) for an opposing view). In addition, osmotic adjustment in both growing and mature regions will serve to sustain the cells in question at a viable water content for an extended period of time and thereby delay cell

Fig. 2. Water potential (▲, △) and osmotic potential (● ○) of the apical 6 mm of nodal roots of maize plants watered daily (– – –) or not watered after day 0 (——). This region includes the meristem and part of the zone of elongation. The difference between the water and osmotic potentials indicates the magnitude of turgor. Points are means ± S.E. ($n = 4$). Modified from Sharp & Davies (1979).

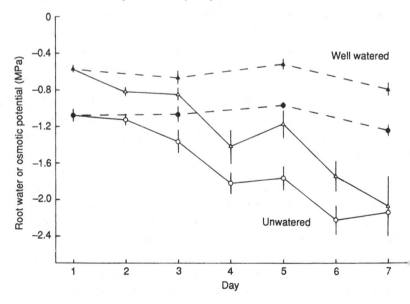

<header>Regulation of growth with restricted water supply</header>

<body>

Fig. 3. Shoot development during the times taken for the primary root of maize to grow approximately 10 cm in vermiculite of various water potentials ($\psi_w$). The treatments represent the vermiculite water contents as percentages of the water content at high water potential. Seedlings were transplanted to the different treatments 30 h after planting, and were grown in the dark at 29 °C and near saturation humidity. The duration of growth after transplanting ranged from 40 h (100% treatment) to 100 h (2% treatment). See Fig. 4 (inset) for corresponding root elongation rates. From Sharp, Silk & Hsiao (1988), with permission.

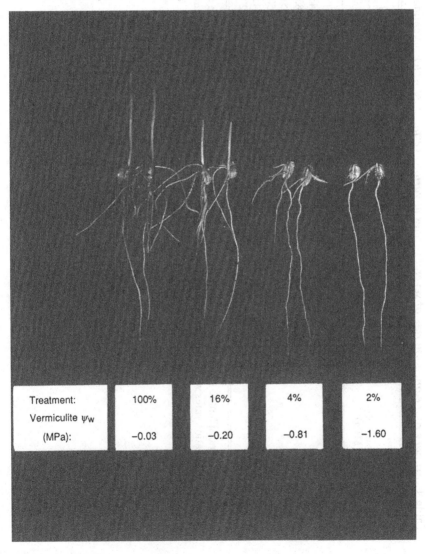

| Treatment: | 100% | 16% | 4% | 2% |
|---|---|---|---|---|
| Vermiculite $\psi_w$ (MPa): | −0.03 | −0.20 | −0.81 | −1.60 |

</body>

Table 1. *Primary root elongation rate of several species at various vermiculite water potentials*

| Vermiculite water potential (MPa) | Primary root elongation rate (mm h$^{-1}$) | | | | |
|---|---|---|---|---|---|
| | Maize | Soybean | Bean | Cowpea | Squash |
| − 0.03 | 2.87±0.41 | 2.34±0.38 | 2.62±0.41 | 2.71±0.52 | 2.37±0.43 |
| − 0.26 | 1.70±0.31 (59%) | 1.82±0.28 (78%) | 1.96±0.19 (75%) | 2.24±0.36 (83%) | 1.69±0.43 (71%) |
| − 0.68 | 1.52±0.31 (53%) | 1.32±0.40 (56%) | 1.46±0.22 (56%) | 1.35±0.30 (50%) | 0.94±0.20 (40%) |
| − 1.18 | 1.11±0.22 (39%) | 1.03±0.23 (44%) | 1.27±0.38 (49%) | 0.74±0.19 (27%) | 0.68±0.16 (30%) |

Seedlings were transplanted to the different water potentials 30 h after planting, and were grown in the dark at 29 °C and near saturation humidity. Elongation rates were constant when the measurements were made. Data of R.E. Sharp and G. Voetberg (unpublished).

death (Flower & Ludlow, 1986). Osmotic adjustment in mature leaves may also help to sustain photosynthesis by maintaining leaf water content at reduced water potentials and this could clearly be advantageous. This has perhaps been demonstrated by the selection of wheat varieties with a higher capacity for osmotic adjustment in mature leaves (Morgan, 1983; Morgan & Condon, 1986). Interestingly, the carbohydrate gained as a result of this increased adjustment leads to deeper rooting and consequently greater water uptake. This study is an excellent example of how yield may be improved under stress conditions by selecting for a particular physiological response, although other factors probably also differed in the different varieties.

### Sustained growth of roots in drying soil

Sharp & Davies (1979) found that nodal roots of maize plants had substantial capacity for osmotic adjustment (Fig. 2), and that this correlated well with continued growth of these roots at low water potentials. The results of Westgate & Boyer (1985a) have confirmed this view (Fig. 1). It seems likely that the successful penetration of even a few nodal axes through dry upper soil layers may be of considerable advantage for the establishment of an adequate root system. In addition, these roots may play an important role in the period immediately after dry soil is recharged with water (Shone & Flood, 1983). Recent work with primary roots of maize (Sharp, Silk & Hsiao, 1988) also shows substantial rates of root elongation at water potentials lower than $-1.5$ MPa, whereas shoot growth was completely inhibited at $-0.8$ MPa (Fig. 3). Similar results have been obtained in a range of species (Table 1). Continued elongation of primary roots at low water potentials is of obvious importance for seedling establishment because of the vulnerability of surface soil to drying.

The mechanism of osmotic adjustment in roots growing at low water potential is therefore an important process to understand. Using maize seedlings growing in vermiculite at the different water potentials shown in Fig. 3, Sharp and co-workers characterised the spatial distribution of root elongation rate for individual primary roots (Fig. 4a). Slower elongation rates at lower water contents were shown to result from decreases in the length of the growing zone, although local rates of elongation were unaffected close to the root apex. Spatial distributions of osmotic potential (Sharp, Hsiao & Silk, 1989) and soluble carbohydrate within the root growing zone were also determined. This information was then used to evaluate profiles of the deposition rate of total osmoticum and soluble carbohydrate, as described by Silk *et al.* (1986).

Despite large decreases in the osmotic potential profile of the root tip

Fig. 4. Spatial distribution of (*a*) relative elemental elongation rate (longitudinal growth rate) and (*b*) osmotic potential in the apical 10 mm of maize primary roots growing at various vermiculite water contents (see Fig. 3). Growth distributions were obtained by time-lapse photographic analysis of the growth of marked roots; points are means from 5 or 6 roots. Osmotic potentials were measured on bulked samples from 30–50 roots; points are means ± S.D. (*n* = 3–7). Root elongation rates (*a*, inset) were constant when the measurements were made. Modified from Sharp *et al.* (1988, 1989).

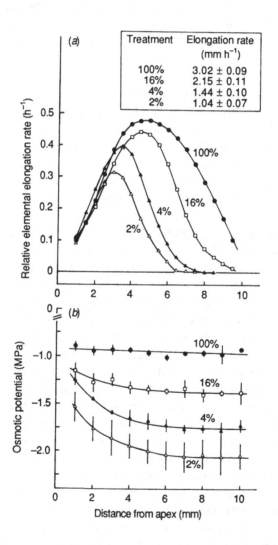

(Fig. 4*b*), osmoticum deposition rates were not increased at any location in any of the low water potential treatments (Fig. 5). Similar results were obtained for soluble carbohydrate. The changes in the profile of osmoticum deposition at low water potentials were very similar to those of root elongation (compare Figs 4*a* and 5). Therefore, decreases in elongation also could not account for the changes in osmotic potential, i.e. solutes did not 'pile up'. The critical response which apparently explains how root tips adjust their osmotic status at low water potential is a decrease in radial expansion (Fig. 6, inset). When expressed volumetrically, rates of osmoticum deposition in the apical region of the roots where elongation growth continued were increased greatly at low water potentials (Fig. 6). These results point towards a regulation of radial expansion as a means of sustaining turgor and longitudinal growth in drying soil. This is clearly a very different mechanism of osmotic adjustment to that described above for leaves and stems, and it will be interesting to perform this type of analysis

Fig. 5. Spatial distribution of net osmoticum deposition rate per mm root length in the apical 10 mm of maize primary roots growing at various vermiculite water contents (see Fig. 3). The data were computed from distributions of growth rate and osmotic potential (Fig. 4), as described by Silk *et al.* (1986). The inset shows the total osmoticum deposition rate in the apical 9 mm for the different treatments, calculated by integrating the rates over distance. Modified from Sharp *et al.* (1989).

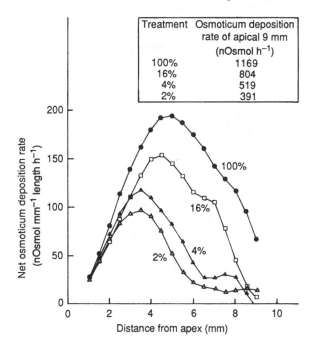

with nodal roots, the elongation rate of which apparently shows even less sensitivity to low water potentials (Fig. 1). Further investigations are required to quantify the anatomical basis of the response and to identify the control system involved.

In primary roots growing at low water potentials, osmotic potentials were decreased throughout the root tip but less so close to the apex (Fig. 4b). These results suggest that turgor was probably higher in the basal regions where elongation was inhibited (Fig. 4a) than in the apical region where elongation was unaffected. It seems likely, therefore, that the inhibition of root elongation at low water potentials was attributable to some factor other than lack of turgor. Pritchard, Tomos & Wyn Jones (1987) have reached

Fig. 6. Spatial distribution of net osmoticum deposition rate per mm$^3$ of tissue water in the apical 10 mm of maize primary roots growing at various vemiculite water contents (see Fig. 3). The data were obtained by dividing rates per mm length (Fig. 5) by the volume of water in each segment. The inset shows root diameter as a function of distance from the apex in the different treatments. Points are means ± s.D. ($n = 5$–6). Modified from Sharp *et al.* (1988, 1989).

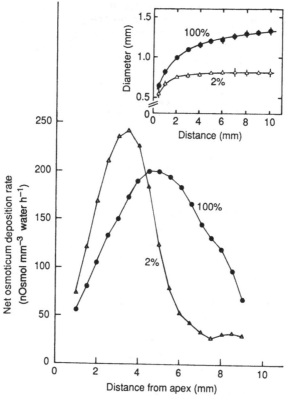

similar conclusions in their work on the growth limitation of 'low-salt' wheat roots.

Restriction of root radial expansion at low water potentials is an effective way of efficiently exploring soil for water at minimum cost. A similar response to that of maize occurs in the primary root of a range of species (R.E. Sharp & G. Voetberg, unpublished), and was noted previously in cotton and peanut by Taylor & Ratliff (1969), although only at low soil strengths. Mechanical impedance to root penetration increases as soil dries and commonly causes roots to become thicker as their elongation rate is reduced (e.g. Wilson, Robards & Goss, 1977), and this may confound direct effects of low water potential that might otherwise occur.

### Some effects of increased mechanical impedance of soil on roots and shoots

To this point we have emphasised that soil drying will reduce the supply of water to plants and that this can have substantial effects on

Fig. 7. Root and shoot dry weight of wheat after 22 days of growth (5-leaf stage) at various soil penetrometer resistances. Variations in penetrometer resistance were obtained by varying soil bulk density and water content. Symbols are as follows. Shape refers to bulk density (g cm$^{-3}$): O, 1.17; $\triangle$, 1.29; $\square$, 1.37; $\diamondsuit$, 1.41; $\nabla$, 1.45. Shade refers to water content (g g$^{-1}$ dry soil): open symbols, 0.22 or 0.23; half-shaded, 0.25; closed, 0.27. Points are means ± s.e. ($n = 6$). Modified from Masle & Passioura (1988).

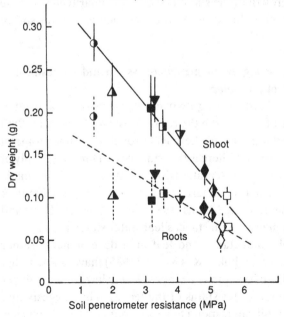

growth. As noted, however, the mechanical impedance of soil increases as soil dries and in itself this can greatly reduce the growth of both roots and shoots (Taylor & Ratliff, 1969; Masle & Passioura, 1988). Somewhat surprisingly, effects of increasing mechanical impedance can be greater on shoots than on roots (Fig. 7). In a recent experiment where wheat seeds were placed on soil packed to different bulk densities and covered with loose soil, leaf expansion was reduced in plants in soil with high impedance before the first leaf was fully expanded (Masle & Passioura, 1988). Relative rates of leaf expansion thereafter were consistently lower in plants in soil at high mechanical impedance, and stomatal conductances were also reduced. Effects on leaf growth and stomatal conductance were the same whether variations in impedance were brought about by changes in soil water content or in bulk density (Fig. 7).

While it has been suggested that roots grow poorly in compacted soil because they are unable to build up sufficient turgor to push aside the soil (Greacen & Oh, 1972), the causes of the reduction in leaf growth are not well understood. Masle & Passioura (1988) list three possible causes of reduced shoot growth as mechanical impedance of soil increases: (1) a limiting supply of nutrients or (2) a limiting supply of water, both as a result of a restricted root system, and (3) a reduced carbon supply because of a higher carbon demand from the roots or because of low stomatal conductance. Their experiments led them to rule out all three possibilities, and they speculate that the growth of the shoot is reduced primarily in response to some hormonal message induced in the roots as they encounter soil of high mechanical impedance.

### 'Sensing' of soil drying by the plant root system and the resulting regulation of shoot physiology

We have discussed how shoot growth can be restricted as a result of a reduction in the supply of water from the roots. Masle & Passioura (1988) have, however, described a situation where leaf growth rates and stomatal conductance are low even though there is no evidence of a limiting supply of water. In recent years several observations of this type have been made (e.g. Bates & Hall, 1981; Davies & Sharp, 1981; Jones, 1983; Blackman & Davies, 1985; Gollan, Passioura & Munns, 1986), leading Jones (1985) and others to suggest that in many circumstances leaf water status is controlled by variation in stomatal conductance and leaf area development, rather than the converse. Turner, Schulze & Gollan (1985) have shown that stomatal conductance of sunflower may be more closely linked to the water status of the soil than to the water status of the leaves. This suggests that shoots of plants in drying soil can respond to a root-sourced 'signal' which is independent of any hydraulic effect. It seems possible that shoot growth and development might be influenced more by a signal which somehow indicates gradual changes in soil water availability than by, for example, comparatively short-term fluctuations in leaf turgor.

### The nature of a root-sourced chemical signal

The nature of a root-sourced signal of soil drying has been the subject of much debate (see e.g. Davies *et al.*, 1986; Schulze, 1986) and considerable recent experimentation. Early demonstrations that the cytokinin supply from roots may be reduced by soil drying treatments (e.g. Itai & Vaadia, 1965) have been supported by more recent evidence (Blackman & Davies, 1985), but it is difficult to see how such a signal could be sensitive enough to relay information on changes in soil water content to the shoots with any precision (see Davies *et al.*, 1987). While soil drying may initially reduce cytokinin transport by only a few per cent, the log-linear relationship between, for example, stomatal aperture and the concentration of cytokinin applied suggests that small changes in the magnitude of such a chemical signal should have negligible effects.

It is known that cytokinin can interact with abscisic acid (ABA) to influence stomata (e.g. Radin, Parker & Guinn, 1982; Blackman & Davies, 1983), and that inorganic nutrients can also change the sensitivity of stomata to ABA (Radin *et al.*, 1982). Observations of this type suggest that any change in the functioning of roots might influence shoot physiology via any number of chemical effects. For example, if soil drying reduces root activity, the uptake of nitrogen and phosphorus may be reduced and cytokinin synthesis in root tips may also be restricted. All of these changes would be expected to enhance the sensitivity of stomata to any ABA that is present in the leaf. We know that even leaves of well-watered plants can contain substantial amounts of ABA, but that in the light this is sequestered within the chloroplasts (e.g. Hartung, Gimmler & Heilmann, 1982). Presumably, then, there would have to be some redistribution of ABA within the leaf before such a signal would influence stomata. ABA redistribution can occur as a result of perturbation of the pH relations of the leaf (Heilmann, Hartung & Gimmler, 1980). Recent experiments by Hartung, Radin & Hendrix (1988) show that dehydration of cotton leaves in a pressure chamber caused large increases in the apoplastic pH. This in turn released ABA from the mesophyll cells into the apoplastic fluid. It seems possible that dehydration of root cells in drying soil could lead to a reduction in pH of xylem sap which might stimulate redistribution of ABA within a turgid leaf and thereby promote stomatal closure.

After some early controversy, it is now clear that roots also have the capacity to synthesise ABA (Walton, Harrison & Cote, 1976; Cornish & Zeevaart, 1985; Lachno & Baker, 1986). Recent work by Zhang & Davies (1987) shows that only slight dehydration of isolated root tips stimulates a substantial buildup of ABA (Fig. 8), and that ABA synthesised in root tips in response to soil drying can move to the leaves in the transpiration stream, a proportion of which arrives at the evaporating sites in the epidermis (Meidner, 1975). We have noted already that ABA in the leaves of well-watered plants is sequestered in the chloroplasts. Therefore, root-sourced ABA can act quite independently of any ABA which is already present in the leaf.

This mechanism might provide a very sensitive means of detecting small changes in the water status of the soil in the immediate vicinity of each individual root tip. Indeed, comparatively few roots might generate a signal which would influence shoot physiology even though the bulk of the root system was in soil with a high water content. A demonstration of this type of response has been provided by Zhang, Schurr & Davies (1987). *Commelina*

Fig. 8. ABA accumulation in detached root tops (apical 20–30 mm) of *Commelina* as a function of root tip turgor. Root tips were excised from well-watered plants and dried in air at 23 °C in the dark until different percentages of fresh weight had been lost. This process took between 5 and 20 min. The samples were then maintained at the various water contents for 7 h prior to measurements of water relations and ABA content. Points are means ± s.e. of at least four measurements. Modified from Zhang & Davies (1987).

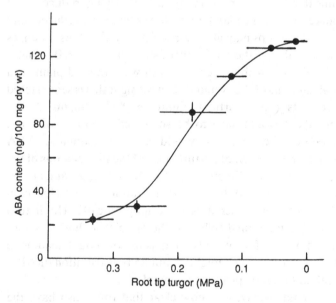

plants were grown with roots split between two pots and after a period where both halves of the root system were watered well, water was withheld from one pot. Roots in this pot soon dried the soil and stomatal conductance was reduced (Fig. 9) even though leaf water potential and turgor of the half-watered plants were no lower than those of plants which had water applied regularly to both halves of the root system. There is therefore no evidence that in this experiment leaf conductance was restricted by limited water supply to the leaves. Stomata apparently closed because of an

Fig. 9. Leaf water potential and turgor, abaxial stomatal conductance, and ABA content of abaxial epidermis of leaves of *Commelina* plants which were grown with their root systems divided between two pots. Water was either applied daily to both halves of the root system (▲) or was withheld from one half of the root system after day 1 of the experimental period (△). Points for water relations and conductance are means ± S.E. Modified from Zhang, Schurr & Davies (1987).

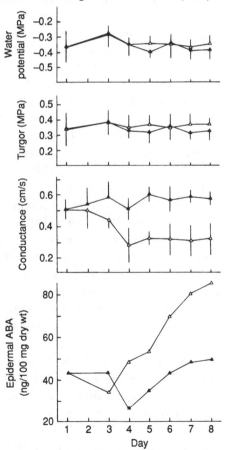

increase in epidermal ABA content (Fig. 9), and there is good evidence that this ABA originated in the roots in drying soil.

ABA clearly has a potent effect on stomata but it is also well known as a growth inhibitor and it influences many aspects of plant development (e.g. Quarrie & Jones, 1977; Quarrie, 1984). Van Volkenburgh & Davies (1983) showed that ABA can reduce the extensibility of the walls of growing leaf cells and it is therefore possible that root-sourced ABA could also play a direct role in the regulation of growth and development of plants growing in drying soil, and may perhaps contribute to the inhibition of leaf and stem growth that occurs at low water potentials despite the maintenance of turgor in their growing regions.

### *'Measurement' of soil moisture status*

In order to develop further ideas on root to shoot communication of information on the water status of the soil, it is necessary to understand how the plant might 'measure' a change in the water status of the root zone and how this measurement might be translated into a regulation of shoot physiology. It seems possible that a simple dehydration of an increasing number of root tips as the soil dries will stimulate increasing ABA synthesis in the roots and transport to the leaves. Measurements of root water relations are few and far between, but Sharp & Davies (1979) measured the water relations profile along nodal roots of maize plants growing in drying soil and noted that root tips were rather isolated hydraulically, perhaps because of only limited xylem development in this region. A recent experiment by Zhang & Davies (1989) seems to have confirmed this view. Maize plants were established in deep tubes of soil which were then not watered for a period of 21 d. There was a substantial decrease in water potential and turgor of the apices of fine secondary roots in the upper soil layers early in the experimental period (Fig. 10). Leaf water relations were apparently unaffected at this time. In contrast, nodal and seminal roots in the same regions of the soil profile retained turgor for the whole of the experimental period, as did both small and large roots deeper in the soil. These results demonstrate that different classes of roots may react very differently to soil drying, and that the root system cannot be considered as a homogeneous mass.

Interestingly, stomatal conductance was substantially restricted some two weeks before leaf water potential declined, again suggesting some non-hydraulic influence of the drying soil on shoot physiology. Thus, the partial dehydration of shallow, possibly non-growing secondary roots may provide a sensitive means of detecting the degree of soil drying. Despite this, there may be sufficient root growth in other parts of the soil profile to sustain substantial water flux to leaves, as shown by Sharp & Davies (1985). Indeed, in the experiment described by Zhang & Davies (1989) root dry weight increment was actually enhanced by soil drying. It may be that increased carbohydrate was made available as a result of a limitation in shoot development induced by soil drying, again via the influence of a chemical signal from the roots. Because the shoot/root ratio of maize seedlings is so high, a small inhibition of shoot growth may allow a substantial increase in root growth.

Fig. 10. Leaf water potential and abaxial stomatal conductance (upper figure), and water potential and turgor of secondary and tertiary root tips (lower figure) of maize plants growing in 1 m deep soil columns, watered daily (▲) or not watered after day 0 (△). The roots were sampled from the upper 20 cm of the soil column. Plants were 20 days old at the beginning of the experimental period. Points are means ± s.e. Modified from Zhang & Davies (1989).

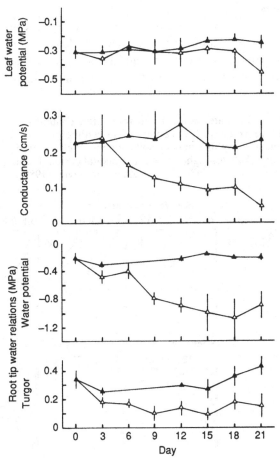

Roots of well-watered maize plants generally contain less than 4 ng ABA/100 mg root dry weight. Even mild soil drying stimulated ABA accumulation in roots in shallow soil (Fig. 11), and increasing amounts of ABA were produced as soil water content decreased. This had the effect of generating a 'front' of ABA accumulation which moved deeper and deeper in the soil with time from last watering. From the results of this experiment it is apparent that the metabolism of individual roots can be relatively independent. Each will contribute to the chemical signal moving through the xylem to the shoot. Roots in very dry soil apparently contained reduced amounts of ABA but this may have been because many of the roots in this soil were dead. In soil of intermediate dryness (soil water content greater than $0.1$ g cm$^{-3}$), it is clear that root ABA content provides a sensitive indicator of soil water status around the roots (Fig. 12).

### Conclusions

We have noted how plant growth and development can be influenced by limited water availability in the soil. Direct effects of limited water supply are rather different in the leaves and the roots and it seems particularly important that continued growth of nodal and primary roots is

Fig. 11. Bulk soil water content and root ABA content in consecutive 10 cm soil layers on days 3, 9, 12 and 18 after withholding water ($\triangle$) from maize plants growing in 1 m deep soil columns. Well-watered plants ($\blacktriangle$) received water daily throughout the experimental period. Points are means of five measurements. Modified from Zhang & Davies (1989).

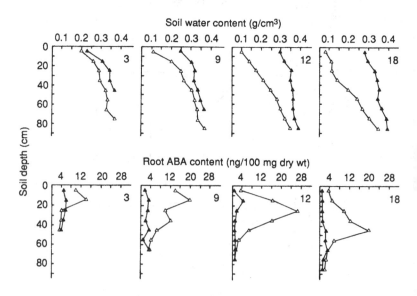

Fig. 12. Relationships between root ABA content and bulk soil water content for maize plants growing in drying soil columns. Data are from Fig. 11, but do not include soil water contents less than 0.1 g cm$^{-3}$ in which many roots were non-living. Modified from Zhang & Davies (1989).

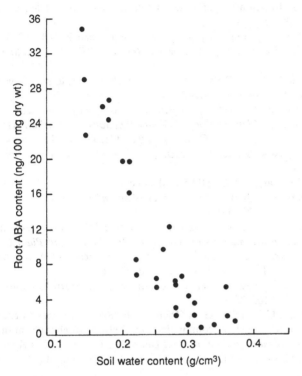

possible in drying soil as a result of turgor maintenance by osmotic adjustment. Interestingly, secondary and tertiary roots may have a controlling-influence on shoot growth and development via the effect of a chemical signal generated in dehydrating roots in drying soil. This may occur when there is no obvious hydraulic effect of soil drying on shoots. In some cases, presumably, hydraulic and chemical effects will be felt together. A chemical signal of the type discussed above may enable the plant to regulate growth and development as a function of the amount of water available in the soil. Jones (1980) has calculated the advantages of this type of regulation of stomata set against constitutive or non-responsive regulation. The advantages in terms of carbon gain and chances of survival can be substantial.

### References

Bates, L.M. & Hall, A.E. (1981). Stomatal closure with soil water depletion not associated with changes in bulk leaf water status. *Oecologia*, **50**, 62–5.
Blackman, P.G. & Davies, W.J. (1983). The effects of cytokinins and ABA on

stomatal behaviour of maize and *Commelina*. *Journal of Experimental Botany*, **34**, 1619–26.

Blackman, P.G. & Davies, W.J. (1985). Root to shoot communication in maize plants of the effects of soil drying. *Journal of Experimental Botany*, **36**, 39–48.

Boyer, J.S. (1968). Relationship of water potential to growth of leaves. *Plant Physiology*, **43**, 1056–62.

Boyer, J.S. (1982). Plant productivity and environment. *Science*, **218**, 443–8.

Boyer, J.S. (1985). Water transport. *Annual Review of Plant Physiology*, **36**, 473–516.

Bozarth, C.S., Mullet, J.E. & Boyer, J.S. (1987). Cell wall proteins at low water potentials. *Plant Physiology*, **85**, 261–7.

Bradford, K.J. & Hsiao, T.C. (1982). Physiological responses to moderate water stress. In *Physiological Plant Ecology II, Water Relations and Carbon Assimilation*, ed. O.L. Lange, P.S. Nobel, C.B. Osmond and H. Ziegler. Encyclopedia of Plant Physiology, New Series, Vol. 12B, pp. 253–232. New York: Springer-Verlag.

Cornish, K. and Zeevaart, J.A.D. (1985). Abscisic acid accumulation by roots of *Xanthium strumarium* L. and *Lycopersicon esculentum* Mill. in relation to water stress. *Plant Physiology*, **79**, 653–8.

Davies, W.J. & Sharp, R.E. (1981). The root: a sensitive detector of a reduction in water availability? In *Mechanisms of Assimilate Distribution and Plant Growth Regulators*, ed. J. Kralovic, pp. 53–67. Prague: Slovak Society of Agriculture.

Davies, W.J., Metcalfe, J., Lodge, T.A. & Rosa da Costa, A. (1986). Plant growth substances and the regulation of growth under drought. *Australian Journal of Plant Physiology*, **13**, 105–25.

Davies, W.J., Schurr, U., Taylor, G. & Zhang, J. (1987). Hormones as chemical signals involved in root to shoot communication of effects of changes in the soil environment. In *Hormone Action in Plant Development – A Critical Appraisal*, ed. G.U. Hoad, M.B. Jackson, J.R. Lenton and R. Atkin, pp. 201–6. London: Butterworths.

Flower, D.J. & Ludlow, M.M. (1986). Contribution of osmotic adjustment to the dehydration tolerance of water-stressed pigeon pea (*Cajanus cajan* (L.) Millsp.) leaves. *Plant, Cell and Environment*, **9**, 33–40.

Gollan, T., Passioura, J.B. & Munns, R. (1986). Soil water status affects the stomatal conductance of fully turgid wheat and sunflower leaves. *Australian Journal of Plant Physiology*, **13**, 1–7.

Greacen, E.L. & Oh, J.S. (1972). Physics of root growth. *Nature New Biology*, **235**, 24–5.

Hartung, W., Gimmler, H. & Heilmann, B. (1982). The compartmentation of abscisic acid, of ABA – biosynthesis, ABA – metabolism and ABA conjugation. In *Plant Growth Substances 1982*, ed. P.F. Wareing, pp. 325–334. London: Academic Press.

Hartung, W., Radin, J.W. & Hendrix, D.L. (1988). Abscisic acid movement into the apoplastic solution of water-stressed cotton leaves: Role of apoplastic pH. *Plant Physiology*, **86**, 908–13.

Heilmann, B., Hartung, W. and Gimmler, H. (1980). The distribution of abscisic acid between chloroplasts and cytoplasm of leaf cells and the permeability of the chloroplast envelope for abscisic acid. *Zeitschrift für Pflanzenphysiologie*, **97**, 67–78.

Herrero, M.P. & Johnson, R.R. (1981). Drought stress and its effect on maize reproductive systems. *Crop Science*, **21**, 105–10.

Itai, C. & Vaadia, Y. (1965). Kinetin-like activity in root exudate of water-stressed sunflower plants. *Physiologia Plantarum*, **18**, 941–4.

Jones, H.G. (1980). Interaction and integration of adaptive responses to water stress: the implications of an unpredictable environment. In *Adaptation of Plants to Water and High Temperature Stress*, ed. N.C. Turner and P.J. Kramer, pp. 353–65. New York: Wiley.

Jones, H.G. (1983). Estimation of an effective soil water potential at the root surface of transpiring plants. *Plant, Cell and Environment*, **6**, 671–4.

Jones, H.G. (1985). Physiological mechanisms involved in the control of leaf water status: Implications for the estimation of tree water status. *Acta Horticulturae*, **171**, 291–6.

Jordan, W.R., Dugas, W.A. & Shouse, P.J. (1983). Strategies for crop improvement for drought-prone regions. In *Plant Production and Management under Drought Conditions*, ed. J.F. Stone and W.O. Willis, pp. 281–99. Amsterdam: Elsevier.

Lachno, D.R. & Baker, D.A. (1986). Stress induction of abscisic acid in maize roots. *Physiologia Plantarum*, **68**, 215–21.

Lockhart, J.A. (1965). An analysis of irreversible plant cell elongation. *Journal of Theoretical Biology*, **8**, 264–75.

Malik, R.S., Dhankar, J.S. & Turner, N.C. (1979). Influence of soil water deficits on root growth of cotton seedlings. *Plant and Soil*, **53**, 109–15.

Masle, J. & Passioura, J.B. (1988). The effect of soil strength on the growth of young wheat plants. *Australian Journal of Plant Physiology*, **14**, 643–56,

Matsuda, K. & Riazi, A. (1981). Stress-induced osmotic adjustment in growing regions of barley leaves. *Plant Physiology*, **68**, 571–6.

Matthews, M.A., Van Volkenburgh, E. & Boyer, J.S. (1984). Acclimation of leaf growth to low water potentials in sunflower. *Plant, Cell and Environment*, **7**, 199–206.

Meidner, H. (1975). Water supply, evaporation and vapour diffusion in leaves. *Journal of Experimental Botany*, **26**, 666–73.

Meyer, R.F. & Boyer, J.S. (1972). Sensitivity of cell division and cell elongation to low water potentials in soybean hypocotyls. *Planta*, **108**, 77–87.

Meyer, R.F. & Boyer, J.S. (1981). Osmoregulation, solute distribution, and growth in soybean seedlings having low water potentials. *Planta*, **151**, 482–9.

Michelena, V.A. & Boyer, J.S. (1982). Complete turgor maintenance at low water potentials in the elongation region of maize leaves. *Plant Physiology*, **69**, 1145–9.

Morgan, J.M. (1983). Osmoregulation as a selection criterion for drought tolerance in wheat. *Australian Journal of Agricultural Research*, **34**, 607–14.

Morgan, J.M. (1984). Osmoregulation and water stress in higher plants. *Annual Review of Plant Physiology*, **35**, 299–319.

Morgan, J.M. & Condon, A.G. (1986). Water use, grain yield, and osmoregulation in wheat. *Australian Journal of Plant Physiology*, **13**, 523–32.

Pritchard, J., Tomos, A.D. & Wyn Jones, R.G. (1987). Control of wheat root elongation growth. I. Effects of ions on growth rate, wall rheology and cell water relations. *Journal of Experimental Botany*, **38**, 948–59.

Quarrie, S.A. (1984). Abscisic acid and drought resistance in crop plants, *British Plant Growth Regulator Group News*, **7**, 1–15.

Quarrie, S.A. & Jones, H.G. (1977). Effects of abscisic acid and water stress on development and morphology of wheat. *Journal of Experimental Botany*, **28**, 192–203.

Radin, J.W., Parker, L.L. & Guinn, G. (1982). Water relations of cotton plants under nitrogen deficiency. V. Environmental control of abscisic acid accumulation and stomatal sensitivity to abscisic acid. *Plant Physiology*, **70**, 1066–70.

Schulze, E.-D. (1986). Carbon dioxide and water vapour exchange in response to drought in the atmosphere and in the soil. *Annual Review of Plant Physiology*, **37**, 247–74.

Sharp, R.E. & Davies, W.J. (1979). Solute regulation and growth by roots and shoots of water-stressed maize plants. *Planta*, 147, 43–9.

Sharp, R.E. & Davies, W.J. (1985). Root growth and water uptake by maize plants in drying soil. *Journal of Experimental Botany*, **36**, 1441–56.

Sharp, R.E., Hsiao, T.C. & Silk, W.K. (1988). Growth of the maize primary root at low water potentials. II. Spatial distribution of osmotic adjustment in the growing zone. *Plant Physiology*, (in press).

Sharp, R.E., Silk, W.K. & Hsiao, T.C. (1988). Growth of the maize primary root at low water potentials. I. Spatial distribution of expansive growth. *Plant Physiology*, **87**, 50–7.

Shone, M.G.T. & Flood, A.V. (1983). Effects of periods of localised water stress on subsequent nutrient uptake by barley roots and their adaptation by osmotic adjustment. *New Phytologist*, **94**, 561–72.

Silk, W.K., Hsiao, T.C., Diedenhoffen, D. & Matson, C. (1986). Spatial distribution of potassium, solutes, and their deposition rates in the growth zone of the primary corn root. *Plant Physiology*, **82**, 853–8.

Steponkus, P.L., Shahan, K.W. & Cutler, J.M. (1982). Osmotic adjustment in rice. In *Drought Resistance in Crops with Emphasis on Rice*, pp. 181–94. Los Banõs, Philippines: International Rice Research Institute.

Taylor, H.M. & Ratliff, L.F. (1969). Root elongation rates of cotton and peanuts as a function of soil strength and soil water content. *Soil Science*, **108**, 113–19.

Tomos, A.D. (1985). The physical limitations of leaf cell expansion. In *Control of Leaf Growth*, ed. N.R. Baker, W.J. Davies and C. Ong, pp. 1–33. Cambridge: Cambridge University Press.

Turner, N.C. & Jones, M.M. (1980). Turgor maintenance by osmotic adjustment: A review and evaluation. In *Adaptation of Plants to Water and High Temperature Stress*, ed. N.C. Turner and P.J. Kramer, pp. 87–103. New York: Academic Press.

Turner, N.C., Schulze, E.-D. & Gollan, T. (1985). The responses of stomata and leaf gas exchange to vapour pressure deficits and soil water content. II. In the mesophytic herbaceous species *Helianthus annuus*. *Oecologia*, **65**, 348–55.

Van Volkenburgh, E. & Davies, W.J. (1983). Inhibition of light-stimulated leaf expansion by ABA. *Journal of Experimental Botany*, **345**, 835–45.

Van Volkenburgh, E. & Boyer, J.S. (1985). Inhibitory effects of water deficit on maize leaf elongation. *Plant Physiology*, **77**, 190–4.

Walton, D.C., Harrison, M.A. & Cote, P. (1976). The effects of water stress on abscisic acid levels and metabolism in roots of *Phaseolus vulgaris* and other plants. *Planta*, **131**, 141–4.

Westgate, M.E. & Boyer, J.S. (1985a). Osmotic adjustment and the inhibition of leaf, root, stem and silk growth at low water potentials in maize. *Planta,* **164,** 540–9.

Westgate, M.E. & Boyer, J.S. (1985b). Carbohydrate reserves and reproductive development at low leaf water potentials in maize. *Crop Science,* **25,** 762–9.

Wilson, A.J., Robards, A.W. & Goss, M.J. (1977). Effects of mechanical impedance on root growth in barley, *Hordeum vulgare* L. Effects on cell development in seminal roots. *Journal of Experimental Botany,* **28,** 1216–27.

Zhang, J. & Davies, W.J. (1987). Increased synthesis of ABA in partially dehydrated root tips and ABA transport from roots to leaves. *Journal of Experimental Botany,* **38,** 2015–23.

Zhang, J. & Davies, W.J. (1989). Abscisic acid produced in dehydrating roots may enable the plant to measure the water status of the soil. *Plant, Cell and Environment,* **12,** 73–81.

Zhang, J., Schurr, U. & Davies, W.J. (1987). Control of stomatal behaviour by abscisic acid which apparently originates in the roots. *Journal of Experimental Botany,* **38,** 1174–81.

R. GARETH WYN JONES AND J. PRITCHARD

# 6 Stresses, membranes and cell walls

## Introduction

There are possibly two major reasons why stress physiology is currently the subject of extensive research. The first is the value to the exploration of basic plant science of perturbations in relatively simple edaphic and climatic factors such as temperature, aeration, external osmotic pressure and specific salts. The second is the recognition within the international community, both in the rich donor countries and in the poor recipients of that aid, that poor and possibly declining environmental conditions and, more crucially, the amplitude and unpredictability of climatic changes and chronic edaphic conditions, are major threats to the welfare of much of the human race. Therefore understanding and exploiting the resistance of some plants to environmental factors such as drought, waterlogging, high and low temperatures, and salinity, are regarded not simply as physiological or ecological problems, but increasingly as important goals of international economic, political and humanitarian significance.

Use and misuse of the term stress has been discussed in detail in Chapter 1. In this review we explore the use of comparative cell physiology and biochemistry, concentrating on the analysis of drought and salinity and the cellular responses to these stresses, as a means of gaining an insight into the underlying processes and of diminishing our dependence on an anthropomorphic concept of stress as deviation from an agronomically-conceived norm.

## Growth, membranes and cell walls

In plants volumetric growth is primarily the result of cell expansion by the development of a large vacuole. Both growth and the mechanical rigidity of tissues require that a substantial turgor pressure is sustained. A theoretical framework for considering cell expansion has been developed by Lockhart (1965) and others and is outlined below as a basis for discussion of the consequences of environmental perturbations. For a more rigorous analysis, see the reviews of Tomos (1987) and Cosgrove (1986).

The water flux into a cell, and hence the volume increase, is driven by the 'effective' water potential difference between the inside and the outside of the plasmalemma. In calculating an effective water potential difference it is necessary to take account of the reflection coefficient, $\sigma$, a measure of the degree of semipermeability of the membrane. The volumetric increase in cell size with attendant water influx can be described by:

$$\frac{1}{V}\frac{dV}{dt} = \frac{A}{V} L_p(\sigma\Delta\pi_{i-e} - P) = L_v(\sigma\Delta\pi_{i-e} - P) \tag{1}$$

where $(1/V)(dV/dt)$ is the relative rate of increase of volume $(V)$, $L_p$ is the hydraulic conductivity (m s$^{-1}$ MPa$^{-1}$) of the cell membrane(s), which can be converted to a volumetric hydraulic conductance ($L_v$, s$^{-1}$ MPa$^{-1}$) by multiplying by the ratio of the cell volume to the cell surface area $(V/A)$, $\Delta\pi_{i-e}$ is the difference between the intracellular (i) and extracellular (e) osmotic pressures (MPa), and $P$ is the cell turgor pressure (MPa). Although this equation is appropriate for a cell in solution, it becomes necessary to include a wall pressure term for most higher plant cells where the water potential falls below zero, either as a result of transpiration or because of the low soil matric potential (Tomos, 1985).

The relative rate of volume increase of single cells is also related to the wall rheological properties by:

$$(1/V)(dV/dt) = \phi(P - Y) \tag{2}$$

where $\phi$ is the wall yielding coefficient or cell wall extensibility (s$^{-1}$ MPa$^{-1}$), which is a measure of the ease with which cells undergo plastic irreversible expansion, and $Y$ is the yield threshold or turgor that must be exceeded before extension occurs (MPa). During steady state expansions where the rate of wall yielding equals the rate of water uptake, Equations 1 and 2 can be equated to give:

$$\frac{1}{V}\frac{dV}{dt} = \frac{\phi L_v}{(\phi + L_v)} (\sigma\Delta\pi_{i-e} - Y) \tag{3}$$

which reduces to Equation 2 when water transport is not limiting (when $L_v \gg \sigma$). On the other hand, when water transport is limiting ($L_v \ll \sigma$), the turgor pressure approaches the yield stress threshold below which no growth will occur.

This simple framework can be used to consider the potential impact of environmental variables on expansion growth. The major environmental variables such as temperature, relative humidity, wind speed, radiation and rainfall all have short- and long-term effects on the water potential gradient through the soil–plant–atmosphere continuum and thus affect turgor pressure generation and, perhaps more importantly, the effective driving force for growth $(P - Y)$. For example, any change in the extracellular water potential $(\psi_e)$, whether it results from changes in the soil water status or from the more rapid changes arising as a consequence of altered transpiration, will alter the intracellular water potential $(\psi_i)$ and hence the value of the turgor pressure, since (Nobel, 1974):

$$\psi_i = P - \pi_i \tag{4}$$

The rate at which a new equilibrium turgor is achieved in response to an altered external water potential depends on the water flow. The half-time $(t_{1/2})$ of the response for a non-growing cell is given by:

$$t_{1/2} = \ln 2/(L_p(\epsilon + \pi_i)) \tag{5}$$

where $\epsilon$ is the volumetric elastic modulus, defined by $\epsilon = V \, dP/dV$. This parameter is a measure of the reversible elasticity of the cell walls and contrasts with extensibility which refers to the irreversible plastic response. A large value of $\epsilon$ indictates a rigid cell wall. For growing tissues, or cells, the half-time of water exchange can be shown to be dependent on $\sigma$ as well as on $\epsilon$ (Cosgrove, 1986):

$$t_{1/2} = \ln 2/(L_v(\epsilon + \pi) + \phi\epsilon) \tag{6}$$

In practice it is difficult to estimate the various parameters involved in these equations as many of them vary depending on whether one is concerned with isolated cells or whole tissues. Furthermore, estimation of wall rheological properties cannot properly be done on fragments of cell walls.

Short half-times in the range 10–20 s are observed largely because $L_p$ values in single cells for the conductance of both plasma and tonoplast membranes in series are low (around $10^{-7}$ m s$^{-1}$ MPa$^{-1}$; see Tomos, 1987). Thus at the cellular level it would appear that membrane resistance to water flux does little to buffer cell turgor pressure against perturbations in external water potential. However, where significant gradients of water potential occur through *tissues*, for example in shoot apices some distance from a xylem vessel, then this situation may be different. Indeed the existence of such water potential gradients across growing tissues (Molz & Boyer, 1978; Silk & Wagner, 1980) leads to the postulate that water transport to the expanding cells may be growth limiting (see Westgate & Boyer, 1984). Since $L_p$ values for single cells are not thought to be growth limiting (see Tomos, 1987), such gradients must be the consequence of tissue architecture. The resistance to water flow through a transcellular pathway is the sum of individual resistances from source to sink and may be strongly influenced by geometry and a limiting membrane area (Jones *et al.*, 1983, 1988).

In some tissues this resistance may be considerable (Tyree & Jarvis, 1982) but in root apical regions is likely to be small (Pritchard, Tomos & Wyn Jones, 1987). Potentially the suberisation of cell walls will influence not only the apoplasmic pathway for water flow but also the effective membrane area for transcellular flow. The measurement of significant water potential gradients in tissues can also be explained by the presence of solutes in the cell wall. Evidence for such a condition was found in growing pea stems (Cosgrove & Cleland, 1983) and would lead to an underestimation of the water potential gradient required to generate the water flux commensurate with sustained growth.

An alternative mechanism of buffering cell turgor pressure might be the presence of highly elastic cell walls (i.e. a low value of $\epsilon$, see Nobel, 1974; Zimmerman, 1978) as this would maximise volume changes, but dampen pressure changes. However, large volume changes have major implications for tissue architecture and potentially for cytoplastic ionic homeostasis unless cytosolic and vacuolar volumes are regulated independently (see Leigh & Wyn Jones, 1984). Nevertheless, the extent to which the various parameters buffer a plant cell against external perturbation in water potential is not fully understood.

A particularly interesting example is the work of Steudle, Smith & Lüttge (1980) on *Kalanchöe diagremontiana* relating pressure and volume change to water storage capacity during the CAM cycle. In these tissues the volume of a cell of very low $\epsilon$ altered with changes in water content while turgor pressure remained relatively constant.

### Osmotic pressure and solute accumulation

The re-establishment of an adequate turgor to sustain the growth rate after a small decrease in external water potential will depend on

(i) sufficient osmotic adjustment to restore $P$

and/or (ii) a decrease in $Y$ to maintain the effective driving force $(P - Y)$

and/or (iii) increase in wall extensibility ($\phi$) so that a similar growth rate can be obtained at a lower effective force.

There is evidence that all three characters may be modulated in different situations but we will first discuss the osmotic component ($\pi$). Further information on osmotic adjustment, especially in relation to root growth, may be found in Chapter 5. Osmoregulation to compensate for a decrease in cell turgor must involve either ion absorption or organic solute synthesis and accumulation or both. Our understanding of the constraints or principles underlying solute accumulation has evolved substantially in the last decade or so. These can be summarised as (i) a high degree of ionic homeostasis and selectivity in the cytoplasm (cytosol) but not in the vacuole and (ii) the accumulation of compatible or benign organic solutes in the cytoplasm where osmotic pressures in excess of about 1 MPa are being generated (Wyn Jones *et al.*, 1977, 1979). [For convenience, solute concentration and hence osmotic pressure is commonly expressed in terms of osmolality, where 1 osmol contains Avogadro's number of osmotically active particles (Nobel, 1974). An osmotic pressure of 1 MPa is approximately equivalent to 410 mosmol $kg^{-1}$ at 20°C.] Crucial to these hypotheses is the concept of solute compartmentation between cytoplasm and vacuole and steep solute gradients across the tonoplast as well as the plasma membrane. However, the vacuolar membrane cannot sustain a significant pressure differential and must be at overall osmotic equilibrium.

As has been reviewed extensively elsewhere (Wyn Jones & Pollard, 1983), the evidence from many eukaryotic cells and eubacteria suggests common ionic and osmotic characteristics in the cytoplasm of cells, especialy a high $K^+$ selectivity and similar ionic strength giving 300–400 mosmol $kg^{-1}$. In saline habitats both $Na^+$ and $Cl^-$ are actively excluded from the cytosolic compartment (Flowers & Läuchli, 1983) (Tables 1 and 2).

Several reasons for this consistency have been advanced. When we originally postulated the applicability of the model to plants (Wyn Jones *et al.*, 1977), drawing on a previous hypothesis of Steinbach (1962) relating to marine invertebrate animals, we suggested that the effects of ions and ionic strength on enzyme stability and activity via lyotropic effects were at the root of the phenomenon. This hypothesis was reiterated by Yancey *et al.* (1982) but they also expanded the hypothesis to accommodate the accumulation of trimethylamine oxide and urea in elasmobranch fish. These ideas are not entirely compelling as the influence of ionic strength and [$K^+$] on enzymes is quite variable and does not explain why the homeostasis is so pervasive. A more persuasive reason, perhaps, is the ionic requirement and sensitivities of translation (protein synthesis) (Lubin & Ennis, 1964; Weber *et al.*, 1977; Wyn Jones & Pollard, 1983). This unity of the ionic and osmotic relations of the cytoplasms of plant and animal cells is of interest for many

Table 1. *Some examples of cytoplasmic and vacuolar concentrations of $K^+$ in plants estimated by a variety of techniques*

| Method | Tissue | Potassium concentration (mol m$^{-3}$) | | Reference |
|---|---|---|---|---|
| | | Cytoplasm | Vacuole | |
| Steady-state tracer efflux | Beet storage root | 58–86 | 85–205 | Pitman (1963) |
| | Barley roots | 102 | 74 | Pitman & Saddler (1967) |
| | Oat coleoptile | 105–205 | 160–180 | Pierce & Higinbothom (1970) |
| | Onion root | 100 | | Macklon (1975) |
| X-ray micro-probe | 'High salt' barley roots | 92[a] | 79 | Pitman et al. (1981) |
| | 'Low salt' barley | 71–119[a] | 9–12 | Pitman et al. (1981) |
| | Suaeda leaf | 0–16 | 11–24 | Harvey et al. (1981) |
| | Mesophyll cells | | | |
| | Atriplex shoot apex | c. 200[a] | | Storey et al. (1983) |
| Ion profile analysis | 'Low salt' barley roots | 90–110 | 10–20 | Jeschke & Stelter (1976) |
| Potassium-sensitive electrode | Acer pseudoplatanus | 126 | 50 | Rona et al. (1982) |
| | Suspension cells | | | |
| Tissues of low vacuolation | Daucus carota callus | 110 | | Mott & Steward (1972) |
| | Wheat shoot | 150–200 | | Munns et al. (1979) |

[a]Values for cytoplasm determined with beam partially located in cell wall or vacuole.

Table 2. *Estimated cytoplasmic and extracytoplasmic $K^+$, $Na^+$ and $Cl^-$
concentrations in some eukaryotes*

| Species | 'Cytoplasm'[a] | | | Extra-'cytoplasm'[b] | | | Reference |
|---|---|---|---|---|---|---|---|
| | $K^+$ | $Na^+$ | $Cl^-$ | $K^+$ | $Na^+$ | $Cl^-$ | |
| | (mol m$^{-3}$) | | | (mol m$^{-3}$) | | | |
| Man (red blood cell) | 160 | 10 | 3 | 4 | 142 | 101 | a |
| Frog (*Rana*) | 126 | 11 | 10 | 3 | 104 | 74 | b |
| Elasmobranch fish | 187 | 30 | – | 7 | 255 | 241 | b |
| *Mytilus edulis* | 158 | 73 | 56 | 13 | 490 | 573 | b |
| *Nitella edulis* | 101 | 9 | 31 | 116 | 4 | 213 | c |
| *Chara australis* | 112 | 3 | 21 | 112 | 28 | 161 | c |
| Carrot callus (young) | 110 | 20 | 11 | | | | d |
| Oat coleoptile | 180 | 15 | 76 | | | | e |

[a]Cytoplasm, or muscle in case of animals.
[b]Vacuole or plasma or serum.
*Sources:* (a) Pitts, 1974; (b) Prosser, 1973; (c) Tazawa, Kishimoto & Kikuyama,
1974; (d) Mott & Steward, 1972; (e) Pierce & Higinbothom, 1970.

diverse reasons including, in the context of the paper, the interpretation of
'stress' in plants.

In a well-watered, well-fertilised, rapidly growing wheat root, the cytoplasmic ion relations are remarkably similar to those of a human red blood
cell (see Table 2). A wheat cortical cell with the 'normal' ionic and osmotic
characteristics of a eukaryotic cell achieves water potential equilibrium with
a dilute bathing solution (a few mol m$^{-3}$) by the generation of a large
hydrostatic pressure ( = 0.65–0.70 MPa = 300 mosmol kg$^{-1}$) constrained
by a rigid cell wall. On the other hand the red blood cell having no rigid wall
must be bathed in isotonic saline (0.9 NaCl = 300 mosmol kg$^{-1}$). Thus
despite the fundamental differences in cell architecture, function and
growth habit, the cytosolic ionic relations bear a close similarity. Many
plants including wheat are highly sensitive to exposure to 150 mol m$^{-3}$ (*c.*
0.9%) NaCl, so mammals, including man, can perhaps be regarded as
'mobile halophytes'. The lack of a requirement to generate turgor pressure
is of course crucial; to maintain turgor pressure homeostatically at about
0.65 MPa, a wheat root cell bathed in 150 mol m$^{-3}$ NaCl will require to
generate an intracellular osmolality of about 600 mosmol kg$^{-1}$ and thereby
deviate significantly from the eukaryotic 'norm'.

We can speculate from this viewpoint that water stress involves the
generation of an internal osmotic pressure greater than that accommodated

by the normal ion (osmotic) status of the cytosol of a vegetative eukaryotic cell (i.e. *c.* 300–400 mosmol $kg^{-1}$); that is a significant deviation from the 'evolutionary norm'. However, it must be *emphatically* stated that such a 'stress' is part of the normal life cycle of all plants as dehydration frequently occurs during the reproductive phase, for example in seed or pollen formation. Furthermore, in the natural environment, some perturbations in water potential must be experienced because, even at high relative humidities of about 98% which are thermodynamically equivalent to water potentials of about $-2.5$ MPa (see Nobel, 1974), there is a steep gradient of free energy favouring water movement from the leaf to the vapour phase of the atmosphere.

The evolution of structures and mechanisms in plants to regulate water fluxes down these steep thermodynamic gradients and yet maintain the cellular conditions for biochemical activity was a major factor in the colonisation of the terrestrial habitat. Paradoxically, therefore, some 'water stress' is completely 'normal', though some plants are better than others at accommodating large deviations.

The adaptive mechanisms found in drought or salt-resistant plants involve morphological, anatomical and ontogenic as well as cell physiological characteristics such as osmoregulation (see Turner, 1986). At the cellular level the role of compatible solutes in the cytoplasmic phase of osmoregulation has been explored in some detail in halophytes (see Wyn Jones, 1984; Munns *et al.*, 1983). Munns *et al.* (1982) have postulated that the accumulation of these organic solutes in the meristematic tissues can be a limiting factor in certain instances. In the vacuolar phase the dominant characteristic is the flexibility of the range of solutes accumulated in various plants. Yeo (1983) has shown convincingly that the accumulation of organic solutes such as the compatible solutes in the vacuole would generate excessive demands for energy and carbon and potentially nitrogen. However, in an environment where ions are freely available such as a saline habitat, then these ions, $Na^+$, $Cl^-$, etc. can be exploited as vacuolar osmotica at relatively low energy cost although energy must be expended to maintain low cytosolic levels. Where carbon has to be diverted for whole-cell osmoregulation as may occur in xeric habitats then the diversion of energy and skeletal carbon must be a major constraint. In general, possible compartmentation between the so-called compatible cytoplasmic solutes and non-specific vacuolar solutes has not been explored critically in xerophytes. Thus as well as recognising a cytosolic 'evolutionary norm' we must also appreciate that most of the conditions that are referred to as stressful involve additional energy costs both to sustain that homeostatic condition and to allow parallel vacuolar adaptations (see Chapter 1).

Table 3. *Yield threshold and effective turgor pressure in growing tissues*

| Treatment | $Y$ (MPa) | $P - Y$ (MPa) | Reference |
|---|---|---|---|
| Pea stem | 0.3 | 0.2 | Cosgrove, Van Volkenburgh & Cleland (1984) |
| Maize leaf | 0.37 | 0.23 | Hsiao, Silk & Jing (1985) |
| *Nitella* | 0.5 | 0.02 | Green *et al.* (1971) |
| Bean leaves | 0.2–0.4 | 0.2 | Van Volkenburgh & Cleland (1986) |
| Birch leaves | 0.07 | 0.38 | Taylor & Davies (1986) |
| Sycamore leaves | 0.25 | 0.3 | Taylor & Davies (1986) |
| Wheat root | 0.04 | 0.61 | Pritchard (1988) |

**Properties of walls**

As noted earlier, the response of a plant cell to a perturbation in external water potential depends not only on osmotic adjustment but also on wall rheology which may be defined by $Y$ (yield threshold) and $\phi$ (wall extensibility). In a number of studies on shoot tissues, $Y$ has been reported to be large (0.2–0.5 MPa) (Table 3) and in some instances approaching the measured turgor pressure (e.g. *Nitella*: Green, Erickson & Buggy, 1971). Under such circumstances the driving force $(P - Y)$ is significantly less than the actual turgor $(P)$, but a decrease in $Y$ would allow, up to a certain level, the maintenance of growth without *any* turgor adjustment. In this context it is of interest to consider some recent data from this laboratory (Pritchard, 1988). $Y$ in wheat roots was estimated by two methods, one involving short-term changes in growth rate in response to a non-permeant osmoticum and the second the rate of turgor and growth recovery after an osmotic check. In both cases $Y$ was found to be very low (0.05–0.1 MPa) so that the driving force $P - Y$ approximately equalled $P$. It is interesting to speculate that this may be a characteristic of roots. If $Y$ were large and $P - Y$ small, small changes in soil water potentials would be amplified and the driving force for growth would be very sensitive to the micro-heterogeneity of the soil matrix. Growth would be slowed when the root entered a dry area whereas we must assume that roots will need to penetrate, possibly through relatively dry soils, to depths where soil moisture conditions are more favourable. An alternative would be for $Y$ to be modulated very rapidly in response to changes in $(P - Y)$, i.e. homeostatic control of $(P - Y)$ and not $P$ (see Green *et al.*, 1971; Cosgrove, 1987). The role of solute changes, as opposed to rheological changes, in the control of root growth is outlined by Sharp & Davies (Chapter 5).

Table 4. *Wheat root seedling elongation (30 replicates), tip turgor*
*pressure (4 mm, >10 replicates) and tip tensiometric plasticity*
*(2–7 mm, >10 replicates) following various growth altering*
*treatments*

| Treatment[a] | Growth rate (mm/24 h ± s.d.) | Tip turgor pressure (MPa ± s.d.) | Tip plasticity (% extension/ 8 g load ± s.d.) |
|---|---|---|---|
| CaSO$_4$ | 32 ± 0.6 | 0.63 ± 0.02 | 4.7 ± 0.3 |
| NaCl | 30 ± 0.5 | 0.67 ± 0.01 | 4.2 ± 0.2 |
| KCl | 20 ± 0.2 | 0.62 ± 0.01 | 1.3 ± 0.1 |
| ABA | 3.7 ± 1.6 | 0.76 ± 0.06 | 1.1 ± 0.4 |
| Excision | <0.1 | 0.61 ± 0.02 | 0.9 ± 0.1 |

[a]Calcium was present throughout at 0.5 mol m$^{-3}$. NaCl and KCl were
present at 10 mol m$^{-3}$ and ABA at 25 mmol m$^{-3}$ where indicated. For
excision treatment root tips (1 cm) were incubated in 0.5 mol m$^{-3}$ CaCl$_2$.
All measurements were taken after 24 h of the appropriate treatment.

Fig. 1. Decline in tensiometric extensibility of live wheat root tips
(2–7 mm) following excision from the plant. Each point is the mean of 10
determinations performed on methanol-killed tissue.

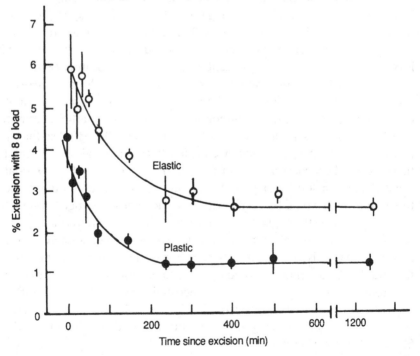

Other studies (Pritchard *et al.*, 1987, unpublished) have shown that the rate of root extension growth may be altered by ions (mainly $K^+$ and $SO_4^{2-}$), ABA and excision. Under all these conditions the turgor pressure of the cells in the elongating zone is strongly homeostatically controlled at about 0.65 MPa and is not correlated with the root elongation rate (Table 4). However, root rheological characters, including in some cases plasticity as measured by an Instron tensiometric method, are correlated with root growth rate, suggesting that these factors affect the wall extensibility $\phi$ (see Pritchard *et al.*, 1987, unpublished). In the case of $K^+$ (Pritchard *et al.*, 1987) and excision (Pritchard *et al.*, unpublished) the changes in wall rheology have been found to be relatively rapid (Figs 1 and 2) with half-times in the $10^3$ s range. Many examples of control of growth by changes in wall rheology have been noted. Water stress was observed to reduce growth rate via an effect on wall rheology in sunflower (Matthews, Van Volkenburgh & Boyer, 1984) and *Phaseolus* leaves (Davies & Van Volkenburgh, 1983). Significantly, Cutler, Rains & Loomis (1977) observed that adaptation to water stress in cotton involved smaller cells and an increase in wall extensibility so that a total osmotic adjustment was not required to maintain growth. It is possible to speculate that

Fig. 2. Decline in tensiometric extensibility of live, intact wheat root tips (2–7 mm) following transfer from 0.5 mol m$^{-3}$ CaCl$_2$ to 10 mol m$^{-3}$ KCl plus 0.5 mol m$^{-3}$ CaCl$_2$. Each point is the mean of 10 determinations performed on methanol-killed tissue.

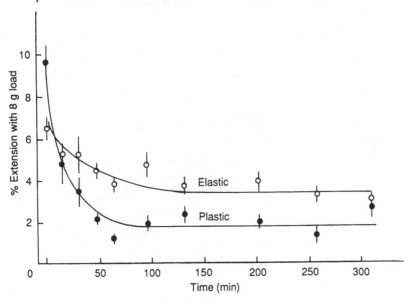

smaller, less rigid cells could allow extension growth at a lower energy cost for osmoregulation. Given the putative energetic cost of organic solute accumulation of about 55 mol ATP per osmol of hexose from fixed $CO_2$, any advantageous change in wall rheology of either $\phi$ or $Y$ would appear highly desirable and of selective advantage. It is becoming increasingly obvious that the regulation of the wall rheological properties is as important as turgor generation and is becoming the subject of more detailed studies both in relation to 'stressed' and 'normal' growth.

### Time base of responses

It is worth considering the time base of the various parameters discussed in relation to turgor pressure regulation or, more properly, the maintenance of an effective turgor pressure for growth (thus considering also changes in $\phi$ and $Y$). Somewhat arbitrarily, Biophysical and Biochemical parameters can be recognised (Fig. 3). It is the pre-existing physical characters, $L_p$ and $\epsilon$, which define the initial half-time of water flow into or out of a cell and thus short-term turgor changes. Longer-term responses (minutes) depend on changes in $\phi$ and $Y$. As noted above, half-times of changes in wheat root plasticity measured tensiometrically following $K^+$ addition and excision are of the order of tens of minutes, though Green *et al.* (1971) reported that $Y$ can alter rapidly following changes in $P$ to restore $(P - Y)$.

Fig. 3. Time course of cellular responses to turgor perturbation.

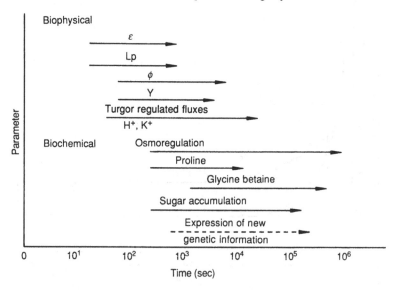

The time base illustrated in Fig. 3 reinforces the dynamic nature of these properties and the importance of their regulation and modulation. The other biophysical aspect is turgor-regulated fluxes. These have been reported in a number of individual cases, e.g. $K^+$ transport in *Valonia* (Zimmermann, Steudle & Lelkes, 1976), phloem loading (Smith & Milburn, 1980) and sucrose uptake in beetroot storage cells (Wyse, Zamski & Tomos, 1986). In the latter example, regulation is associated with turgor-modulated $H^+$ fluxes. However, the control of major ion fluxes such as $K^+$ or $H^+$ in plant cells by turgor has not been widely reported (cf. Cram, 1983). It is likely that the control of fluxes from cytosol to vacuole and from vacuole to cytosol is also involved in this time scale. Any concept of cytoplasmic homeostasis requires the close integration of plasma membrane and tonoplast fluxes as the latter membrane can not sustain a hydrostatic gradient. It is also attractive to postulate that the vacuolar pools of

Fig. 4. Whole-plant fresh weight and leaf osmotic adjustment of *Thinopyrum bessarabicum* as a function of time following the gradual addition of 250 mol m$^{-3}$ NaCl to the culture solution. Fresh weight: ● = control, ○ = 250 mol m$^{-3}$ NaCl. Leaf sap osmotic pressure: □ = 250 mol m$^{-3}$ NaCl.

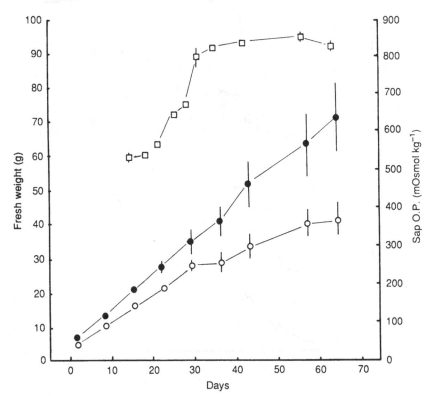

compatible solutes such as proline and betaine can be utilised as cytoplasmic osmotica by activating exchange fluxes before the induction of the *de novo* biosynthesis of these solutes (see Pälich, Jaeger & Horz, 1982).

### An example at the whole-plant level

So far this presentation has dealt exclusively with responses to water potential perturbations at the cellular level. In concluding we wish to consider one example of a response at the whole-plant level in the light of this analysis. This example is developed further by Yeo & Flowers (Chapter 12) who also refer to further examples of the complexity of whole-plant responses.

This laboratory has been involved in studies on the physiological basis of

Fig. 5. Increase in the contents of the major inorganic ions in leaves of *Thinopyrum bessarabicum* as a function of time following the gradual addition of 250 mol m$^{-3}$ NaCl to the culture solution. The continuous line represents the level of the NaCl in the external solution. $\square$ = K, $\bullet$ = Cl, $\bigcirc$ = Na.

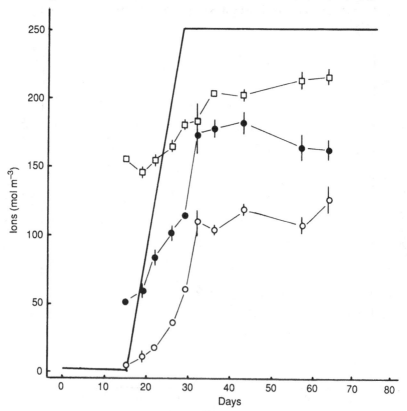

Table 5. *Solute contents of youngest emerged leaves of the amphidiploid, its parent and cultivar Ciano 79 grown for 30 days in non-saline and saline medium[a]*

| | Salt treatment (mol m⁻³) | *Triticum aestivum* cv. Chinese Spring | Amphidiploid[b] | *Triticum aestivum* cv. Ciano 79 | *Thinopyrum bessarabicum[c]* |
|---|---|---|---|---|---|
| Sap osmotic pressure (mosmol kg⁻¹) | 0 | 517±7 | 500±11 | 481±11 | 652±4 |
| | 250 | 855±39 | 634±6 | 725±67 | 786±14 |
| $\Delta\pi$ (0–250) | – | 338 | 134 | 244 | 114 |
| Na (mol m⁻³) | 0 | 20±4 | 28±4 | 20±4 | 21±3 |
| | 250 | 78±15 | 46±1 | 125±38 | 145±65 |
| K (mol m⁻³) | 0 | 198±11 | 184±1 | 181±2 | 190±3 |
| | 250 | 194±5 | 212±1 | 154±17 | 221±2 |
| Cl (mol m⁻³) | 0 | 22±6 | 37±2 | 47±4 | 58±1 |
| | 250 | 151±26 | 85±9 | 194±39 | 149±5 |
| Betaines | 0 | 5.9±0.5 | 3.6±0.2 | 5.1±0.4 | 50±1.0 |
| | 250 | 18.2±0.8 | 19.7±0.9 | 6.5±0.5 | 57±1.0 |

[a] All values are the means of 3 replicates±s.E.
[b] Amphidiploid *Triticum aestivum* cv. Chinese Spring × *Thinopyrum bessarabicum*.
[c] Leaves of the perennial are not strictly comparable with those of the annuals.

the salt tolerance of wheat and related wheat grasses (Gorham *et al.*, 1986, 1987). We will restrict discussion to two pertinent points on the regulation of ion accumulation and osmotic adjustment in the wild wheat grass *Thinopyrum bessarabicum*, the amphidiploid of that grass, and the hexaploid wheat Chinese Spring. Figures 4 and 5 show the changes in leaf sap osmotic pressure and in $K^+$, $Na^+$, and $Cl^-$ concentrations in *T. bessarabicum* following a large salt treatment from 0 to 250 mol m$^{-3}$. This grass, unlike the hexaploid wheats, is able to grow for extended periods at this salinity. Interestingly the osmotic adjustment during the salinisation which illustrates the *potential* for turgor regulation is less than complete, i.e. $\Delta\pi$ (leaf sap − external medium) is less at the high salinity (Fig. 4). While $Na^+$ and $Cl^-$ are absorbed, the fluxes seem tightly regulated and appear to be closed down after the initial phase of partial osmotic adjustment (Fig. 5). In other more salt-sensitive wheat grasses, e.g. *Aegilops umbellulata*, $\Delta\pi$ sap increases after even a low salinity treatment. $Na^+$ and $Cl^-$ influxes are much greater and continue after the phase of initial adjustment. Nevertheless the plants die.

When data on the amphidiploid and the hexaploid wheats are considered (Table 5), the former being substantially more salt tolerant in hydroponic culture, then again several points are noted. First, the sap osmotic pressure in the young leaves is *least* in the tolerant amphidiploid. Secondly, $Na^+$ and $Cl^-$ levels are also lower in the juvenile amphidiploid leaves. These data imply that minimal osmotic adjustment is more beneficial than apparently complete osmotic adjustment.

Since growth is maintained in the amphidiploid and in the wild wheat *T. bessarabicum*, we can only assume that $P - Y$ is adequate to sustain growth at the level of $\phi$ pertaining in this tissue despite the decrease in $\Delta\pi$. In most work on salt adaptation these parameters have not been considered, largely because of the technical problems involved. The effective turgor depends on the membrane reflection coefficient and on $\Delta\pi$ between the vacuolar sap and the cell wall solution (Equation 1). Solute accumulation in the cell wall will cause a $\Delta\pi$ measured as the difference between vacuole and external medium seriously to overestimate turgor. It is clear that in the halophyte *Suaeda maritima* the wall solute concentration is very large (see Clipson *et al.*, 1985). Evidence for the Oertli hypothesis (Oertli, 1968) that salt accumulation in the wall is a major factor in salt toxicity was recently considered by Flowers & Yeo (1986) and the buildup of ions in the wall may well be a highly significant phenomenon. An additional problem that requires examination is the effect of high ion concentration in the wall on $\sigma$, the reflection coefficient. An increase in membrane permeability or a decrease in reflection coefficient, for example by displacement of $Ca^{2+}$, would decrease the effective $\sigma$ and therefore $P$. Also it would increase the energy required to sustain ion gradient as the standing gradient is a function of both the active pump and the passive back leak. An increased permeability would also result in elevated levels of potentially toxic ions in the cytosol. The problem of high solute permeability ($\sigma$ significantly below 1) may also exclude the use of glycerol as a compatible solute in higher plants, while even in fungi glycerol leakage is a problem (see Brown, 1978). Although a *three*-carbon polyhydric alcohol might allow more osmotic adjustment per unit of $CO_2$ fixed than hexose sugar or sugar alcohol accumulation, this has not been found to our knowledge in higher plants.

### Conclusion

This admittedly largely theoretical analysis of the factors affecting the readjustment of growth following a perturbation of external water potential, especially by ions, underlines the close integration of wall and membrane characteristics and undermines the conventional distinction between water stress and specific ion stress. Our inability satisfactorily to determine which is the most damaging stress probably arises from an inadequate analysis of the close integration of the factors at the cellular level. Indeed too often we used the term stress to cover ignorance and an inadequate analysis rather than as an illuminating concept.

### References

Brown, A.D. (1978). Compatible solutes and extreme water stress in eukaryotic microorganisms. *Advances in Microbial Physiology*, **17**, 181–242.

Clipson, N.J.W., Tomos, A.D., Flowers, T.J. & Wyn Jones, R.G. (1985). Salt tolerance in the halophyte *Suaeda maritima* L. Dum. *Planta*, **165**, 393–6.

Cosgrove, D.J. (1986). Biophysical control of plant cell growth. *Annual Review of Plant Physiology*, **37**, 377–405.

Cosgrove, D.J. (1987). Wall relaxation in growing stems: comparison of four species and assessment of measurement techniques. *Planta*, **171**, 266–78.

Cosgrove, D.J. & Cleland, R.E. (1983). Solutes in the free space of growing stem tissues. *Plant Physiology*, **73**, 326–31.

Cosgrove, D.J., Van Volkenburgh, E. & Cleland, R.E. (1984). Stress relaxation of cell walls and the yield threshold for growth: Demonstration and measurement by micropressure probe and psychrometer techniques. *Planta*, **162**, 46–52.

Cram, W.J. (1983). Chloride accumulation as a homeostatic system: set points and perturbations. *Journal of Experimental Botany*, **34**, 74–84.

Cutler, J.M., Rains, D.W. and Loomis, R.S. (1977). The importance of cell size in water relations of plants. *Physiologia Plantarum*, **40**, 255–60.

Davies, W.J. & Van Volkenburgh, E. (1983). The influence of water deficit on the factors controlling the daily pattern of growth of *Phaseolus* trifoliates. *Journal of Experimental Botany*, **34**, 987–99.

Flowers, T.J. & Läuchli, A. (1983). Sodium versus potassium: substitution and compartmentation. In *Inorganic Plant Nutrition, Encyclopedia of Plant Physiology, New Series 15B*, ed. A. Läuchli and R.L. Bieleski, pp. 651–81. Berlin: Springer-Verlag.

Flowers, T.J. & Yeo, A.R. (1986). Ion relations of plants under drought and salinity. *Australian Journal of Plant Physiology*, **13**, 75–91.

Gorham, J., Forster, B.P., Budrewich, E., Wyn Jones, R.G., Miller, T.E. & Law, C.N. (1986). Salt tolerance in the triticaceae – solute accumulation and distribution in an amphidiploid derived from *Triticum aestivum* c.v. Chinese Spring and *Thinopyrum bessarabicum*. *Journal of Experimental Botany*, **37**, 1435–49.

Gorham, J., Hardy, C., Wyn Jones, R.G., Joppa, L.R. & Law, C.N. (1987). Chromosomal location of a $K^+/Na^+$ discrimination character in the D genome of wheat. *Theoretical Abstracts of Genetics*, **74**, 584–8.

Green, P.B., Erickson, R.O. & Buggy, J. (1971). Metabolic and physical control cell elongation rate *in vivo* studies in *Nitella*. *Plant Physiology*, **47**, 423–30.

Harvey, D.M.R., Hall, J.L., Flowers, T.J. & Kent, B. (1981). Quantitative ion localisation within *Suaeda maritima* leaf mesophyll cells. *Planta*, **151**, 555–60.

Hsiao, T.C., Silk, W.K. & Jing, J. (1985). Leaf growth and water deficits: Biophysical effects. In *Control of leaf growth, Society of Experimental Biology Seminar Series 27*, ed. N.R. Baker, W.D. Davies and C. Ong, pp. 239–66. Cambridge: Cambridge University Press.

Jeschke, W.D. & Stelter, W. (1976). Measurement of longitudinal ion profiles in single roots of *Hordeum* and *Atriplex* by use of flameless atomic absorption spectroscopy. *Planta*, **128**, 107–12.

Jones, H., Leigh, R.A., Wyn Jones, R.G. & Tomos, A.D. (1988). The integration of whole root and cellular hydraulic conductivities in cereal roots. *Planta*, **174**, 1–7.

Jones, H., Tomos, A.D., Leigh, R.A. & Wyn Jones, R.G. (1983). Water relation parameters of epidermal and cortical cells in the primary root of *Triticum aestivum* L. *Planta*, **158**, 230–6.

Leigh, R.A. & Wyn Jones, R.G. (1984). A hypothesis relating critical potassium concentrations for growth to the distribution and functions of the ion in the plant cell. *New Phytologist*, **97**, 1–13.

Lockhart, J.A. (1965). An analysis of irreversible plant cell elongation. *Journal of Theoretical Biology*, **8**, 264–75.

Lubin, M. & Ennis, H.L. (1964). On the role of intracellular potassium in protein synthesis. *Biochimica et Biophysica Acta*, **80**, 614–31.

Macklon, A.E.S. (1975). Cortical cell fluxes and transport to the stele in excised root segments of *Allium cepa* L. I. Potassium, sodium and chloride. *Planta*, **122**, 109–30.

Matthews, M.A., Van Volkenburgh, E. & Boyer, J.S. (1984). Acclimation of leaf growth to low water potential in sunflower. *Plant Cell and Environment*, **7**, 199–206.

Molz, F.S. & Boyer, J.S. (1978). Growth induced water potentials in plant cells and tissues. *Plant Physiology*, **62**, 423–9.

Mott, R.L. & Steward, F.C. (1972). Solute accumulation in plant cells. I. Reciprocal relations between electrolytes and ions in carrot explants as they grow. *Annals of Botany*, **36**, 621.

Munns, R., Brady, C.J. & Barlow, E.W.R. (1979). Solute accumulation in the apex and leaves of wheat during water stress. *Australian Journal of Plant Physiology*, **6**, 379–89.

Munns, R., Greenway, H., Delane, R. & Gibbs, J. (1982). Ion concentration and carbohydrate status of the elongating leaf tissue of *Hordeum vulgare* growing at high external NaCl. 2. Cause of the growth reduction. *Journal of Experimental Botany*, **33**, 574–83.

Munns, R., Greenway, H., Stelter, T.L. & Kuo, J. (1983). Turgor pressure, volumetric elastic modulus, osmotic volume and ultrastructure of *Chlorella emersoni* grown at high and low external NaCl. *Journal of Experimental Botany*, **34**, 144–55.

Nobel, P.S. (1974). *Biophysical Plant Physiology and Ecology*. San Francisco: Freeman.

Oertli, J.J. (1968). Extracellular salt accumulation as a possible mechanism of salt injury in plants. *Agosto*, **12**, 461–9.

Pälich, E., Jaeger, H.J. & Horz, M. (1982). Further investigations concerning the energy profile of the glutamic acid proline sequence in water stressed bean plants. *Zeitschrift für Pflanzenphysiologie*, **105**, 475–8.

Pierce, W.S. & Higinbothom, N. (1970). Compartments and fluxes of $K^+$ and $Cl^-$ in *Avena* coleoptile cells. *Plant Physiology*, **46**, 666–73.

Pitman, M.G. (1963). The determination of the salt reactions of the cytoplasmic phase in the cells of beetroot tissue. *Australian Journal of Biological Sciences*, **16**, 647–68.

Pitman, M.G., Läuchli, A. & Stelzer (1981). Ion distribution in roots of barley seedlings as measured by electron probe X-ray micro-analysis. *Plant Physiology*, **66**, 673–9.

Pitman, M.G. & Saddler, H.D.W. (1967). Active sodium and potassium transport in cells of barley roots. *Proceedings of the National Academy of Sciences, USA*, **57**, 44–9.

Pitts, B.J.R. (1974). Relationships of $K^+$ activated phosphatase to $Na^+/K^+$ ATPase. *Annals of the New York Academy of Science*, **242**, 293–304.

Pritchard, J. (1988). The control of growth rate in wheat seedling roots. PhD thesis, University of Wales.

Pritchard, J., Tomos, A.D. & Wyn Jones, R.G. (1987). Control of wheat root elongation growth. I. Effects of ions on growth rate, wall rheology and cell water relations. *Journal of Experimental Botany*, **38**, 948–59.

Prosser, C.L. (1973). *Comparative Animal Physiology*. New York: Saunders.

Rona, J.P., Cornel, D., Gignon, C. & Heller, R. (1982). The electrical potential difference across the tonoplast of *Acer pseudoplantus* cells. *Physiologie Végétale*, **20**, 459–63.

Silk, W.K. & Wagner, K.K. (1980). Growth sustaining water potential distributions in the primary corn root. *Plant Physiology*, **66**, 859–63.

Smith, J.A.C. & Milburn, J.A. (1980). Osmoregulation and the control of phloem sap composition in *Ricinus communis* L. *Planta*, **148**, 28–34.

Steinbach, H.B. (1962). Ionic and water balance of planarians. *Biological Bulletin*, **122**, 310.

Steudle, E., Smith, J.A.C. & Lüttge, U. (1980). Water relations parameters of individual mesophyll cells of the crassulacean acid metabolism plant *Kalanchoe diagremontiana*. *Plant Physiology*, **66**, 1155–63.

Storey, R., Pitman, M.G., Stelzer, R. & Carter, C. (1983). X-ray microanalysis of cells and cell compartments of *Atriplex spongiosa*. I. Leaves. *Journal of Experimental Botany*, **34**, 778–94.

Taylor, G. & Davies, W.J. (1986). Yield turgor of growing leaves of *Betula* and *Acer. New Phytologist*, **104**, 347–53.

Tazawa, M., Kishimoto, U. & Kikuyama, M. (1974). Potassium, sodium and chloride in the protoplasm of *Characea. Plant and Cell Physiology*, **15**, 103–10.

Tomos, A.D. (1985). The physical limitations to leaf cell expansion. In *Control of Leaf Growth*, ed. N.R. Baker, W.J. Davies and C.K. Ong, pp. 1–33. Cambridge: Cambridge University Press.

Tomos, A.D. (1987). Cellular water relations of plants. In *Water Science Reviews, Vol. 3*, ed. F. Franks, pp. 186–227. Cambridge: Cambridge University Press.

Turner, N.C. (1986). Adaptation to water deficits: a changing perspective. *Australian Journal of Plant Physiology*, **13**, 175–90.

Tyree, M.T. & Jarvis, P.G. (1982). Water in tissues and cells. In *Encyclopedia of Plant Physiology, New Series, Vol. 12B*, ed. O.L. Lange, P.S. Nobel, C.B. Osmond and H. Zeigler, pp. 35–77. Berlin: Springer-Verlag.

Van Volkenburgh, E. & Cleland, R.E. (1986). Wall yield threshold and effective turgor of growing bean leaves. *Planta*, **167**, 37–43.

Weber, L.A., Hickey, E.D., Nuss, D.L. & Bayloni, C. (1977). 5′-terminal 7-methyl-guanosine and messenger RNA function influence of potassium concentration on translation *in vitro. Journal of Biological Chemistry*, **252**, 4007–10.

Westgate, M.E. & Boyer, J.S. (1984). Transpiration-induced and growth-induced water potentials in maize. *Plant Physiology*, **74**, 882–9.

Wyn Jones, R.G. (1984). Phytochemical aspects of osmotic adaptation. In *Recent Advances in Phytochemistry 18, Phytochemical Adaptations to Stress*, ed. B.N. Timmermann, K.C. Steelin and F.A. Loewns, pp. 55–78. New York: Plenum Press.

Wyn Jones, R.G., Brady, C.J. & Speirs, J. (1979). Ionic and osmotic relations in plant cells. In *Recent Advances in the Biochemistry of Cereals*, ed. D.L. Laidman and R.G. Wyn Jones, pp. 63–103. London: Academic Press.

Wyn Jones, R.G. & Pollard, A. (1983). Proteins, enzymes and inorganic ions. In *Inorganic Plant Nutrition, Encyclopedia of Plant Physiology, New Series 15*, ed. A. Läuchli and R.L. Bieleski, pp. 528–62. Berlin: Springer-Verlag.

Wyn Jones, R.G., Storey, R., Leigh, R.A., Ahamad, N. & Pollard, A. (1977). A hypothesis of cytoplasmic osmoregulation. In *Regulation of Cell Membrane Activities in Plants*, ed. E. Marre and O. Ciferri, pp. 121–36. Amsterdam: Elsevier/North Holland Biomedical Press.

Wyse, R.E., Zamski, E. & Tomos, A.D. (1986). Turgor regulation of sucrose transport in sugar beet tap root tissue. *Plant Physiology*, **81**, 478–81.

Yancey, P.H., Clark, M.E., Hand, S.C., Bowlus, R.D. & Somero, G.N. (1982). Living with water stress. Evolution of osmolyte systems. *Science*, **217**, 1214–22.

Yeo, A.R. (1983). Salinity resistance; Physiologies and prices. *Physiologia Plantarum*, **58**, 214–22.

Zimmerman, U. (1978). Physics of turgor and osmoregulation. *Annual Review of Plant Physiology*, **29**, 121–48.

Zimmerman, U., Steudle, E. & Lelkes, P.I. (1976). Turgor pressure regulation in *Valonia utricularis*. Effect of cell wall elasticity and auxin. *Plant Physiology*, **58**, 608–13.

G.R. STEWART

# 7 Desiccation injury, anhydrobiosis and survival

## Introduction

Water is, of course, essential for plant growth, but one of the themes of this chapter is that it may not be necessary for plant survival. Although most agronomically important plants are very sensitive to internal water deficits, the majority of plants at some stage of their life cycle are tolerant of desiccation. Few of these have vegetative parts which are desiccation tolerant, but the survival of even so-called drought-evading species, such as the ephemeral desert annuals, rests on the tolerance of their seeds to desiccation.

Desiccation tolerant species may exhibit little or no metabolic activity depending upon the extent of dehydration. In this anhydrobiotic or ametabolic state we are concerned not with metabolic perturbation but with the stability of organelles, membranes and macromolecules in a dehydrated state. However, in the initial period of rehydration, the passage to a metabolically active state poses particular problems if metabolic 'mayhem' is to be avoided.

Knowledge of how organisms avoid irreversible damage from dehydration is obviously of importance in the preservation of seeds and germplasm. It may also further our understanding of the role played by water in maintaining structural and functional integrity of membranes and macromolecules. The controlled metabolic shutdown which occurs as organisms pass into a state of anhydrobiosis may provide valuable insights into the functional significance of the metabolic responses to water deficits in mesophytes.

### Desiccation tolerant plants

There are representatives of desiccation tolerant species amongst all of the major plant divisions. The water content of many bacterial and fungal spores is low (<25%) and they exhibit great tolerance of desiccation (see Ross & Billing, 1957; Bradbury et al., 1981). Desiccation tolerant cyanobacteria are found in a diverse range of drought-prone habitats.

Lichens are poikilohydric and the classical ecophysiological studies by Lange and his colleagues (see e.g. Lange *et al.*, 1975) have characterised the responses of desert species to desiccation. Tolerance of desiccation is related to the prevailing moisture regimes of their natural habitat, and xeric desert species can be dried out and stored for several months with no adverse effects (Lange, 1969). Many xeric lichens cannot survive if kept continuously wet and require alternating periods of wetting and drying (Farrer, 1976). Moreover, maximum photosynthetic activity in several species occurs at water contents less than saturation (Kershaw, 1972; Crittenden & Kershaw, 1978).

In contrast to vascular plants, most bryophytes lack roots and an efficient conducting system and, although some species are confined to wet habitats, many species are remarkably tolerant of drying out. As might be expected, desert bryophytes are extremely tolerant of prolonged desiccation (Scott, 1982). However, desiccation tolerant species are widely distributed and occur in a variety of habitats. Most obligate epiphytic and epilithic species can recover from periods of desiccation in which their water content is reduced to a few per cent (Proctor, 1982). Dry bryophytes in equilibrium with air in sunny weather have water contents of less than 5% and have water potentials of between $-100$ MPa and $-220$ MPa. Laboratory studies have shown that tolerant species can withstand prolonged periods of desiccation at low relative humidities, indeed many survive better at low than at high relative humidities (Dilks & Proctor, 1974).

As a group, vascular plants have vegetative parts which are on the whole intolerant of even partial dehydration; most are killed if allowed to equilibrate with relative humidities of 92–97% (Iljin, 1931). In contrast, many species of non-vascular plants will survive equilibration with 0–5% relative humidities. There are, however, a small number of ferns and flowering plants which can resume normal physiological functions after air drying. There are over 100 species of these so-called resurrection plants, mostly from arid regions of Southern Africa and Australia. They are predominantly pioneer species in the colonisation of rocks, screes and shallow soils (Gaff, 1977). Resurrection plants are more frequently encountered among the ferns and fern allies than among the angiosperms (Gaff, 1971, 1977; Gaff & Latz, 1978). In general, however, angiosperm species are tolerant of more extreme conditions (Gaff, 1977). Air-dried foliage remains viable for several years and that of *Xerophyta squamosa* and *Coleochloa setifera* was uninjured after five years (Gaff, 1977). Survival of air-dried foliage is greatly reduced when stored at relative humidities above 50% (Gaff, 1977). The water potentials of air-dried foliage can be as low as $-400$ MPa. These resurrection plants appear to fall into two groups on the basis of their response to desiccation. In the first group are species which retain their chlorophyll and maintain subcellular organisation in the desiccated state (see Wellburn & Wellburn, 1976). In the second group are species in which chlorophyll is lost and there is extensive breakdown of organelle structure (Hallam & Gaff, 1978; Hallam & Luff, 1980). It is very striking that these extensive alterations in ultrastructure are essential for survival in the desiccated condition (Gaff & Churchill, 1976).

Although the roots and particularly the shoots of most angiosperms are intolerant of desiccation, their seeds and pollen generally undergo a marked decrease in water content as part of their maturation process. At dehiscence, pollen undergoes rapid desiccation and the moisture content can decrease from over 80% to less than 40% (Keijzer, 1983). The water content of some pollen can be reduced further with no adverse effects on viability and drying of pollen is essential for its survival in storage (Binder, Mitchell & Ballantyne, 1974). There are exceptions; pollen of most species of the Gramineae is sensitive to desiccation (see Goss, 1968).

Ripening seeds, except where they are enclosed in fleshy fruit, lose water until their water content is in equilibrium with atmospheric humidity and at this stage they contain 5–20% water. The water contents of some seeds can be reduced still further with no adverse effects on viability, rather this can enhance their survival in the dried state (Roberts, 1973). The longevity of seeds in this anhydrobiotic state can be prodigious, lasting for several hundred years (Priestley & Posthumus, 1982).

### Desiccation injury

Injuries from desiccation are likely to arise from a combination of physical damage to cellular components following water removal, structural and metabolic consequences of increased solute concentration and chemical damage resulting from metabolic disfunction induced by dehydration.

#### *Membrane damage*

The drying protoplast will be subjected to tension as the result of volume contraction and its adherence to the cell wall. Early observations (Steinbrick, 1900) on desiccation tolerant species showed that the protoplasm does not separate from the wall, but rather that it folds and cavities develop in the wall. Where there are thick-walled cells, localised separation of the plasmalemma from the wall may occur. It seems unlikely, however, that rupture of the plasmalemma normally occurs during desiccation. A more subtle form of membrane damage may arise from dehydration-induced conformational changes. Certainly it is relatively easy to demonstrate that dehydrated membranes exhibit a loss of functional integrity when rehydrated. Simon (1974) has documented the extensive literature showing that dehydrated spores, seeds, pollen and shoots leak large amounts of cellular solutes when allowed to rapidly rehydrate. Where dehydration has not caused permanent injury this leakage of solutes is a transient phenomenon, lasting only a few minutes. Leakage from dried organisms has been explained in terms of alterations taking place in membrane systems during dehydration. Simon (1974) concluded that the evidence from biophysical studies was consistent with the idea that below a water content of 20–30%, normal lamellar structure of the membrane was lost and that the phospholipid bilayer was replaced in part by a hexagonal phase. This is largely hydrophobic but has water-filled channels lined by the polar head groups of the phospholipids. In other words 'what was a relatively waterproof phospholipid bilayer becomes a porous structure'. Moreover membrane protein may be displaced from, or within, the membrane as this restructuring occurs. The consequence will be a loss of compartmentation internally and leakage of cellular solutes externally when the dried membrane is rehydrated. Difficulties in demonstrating the presence of a hexagonal phase in the membranes of dried organisms (McKenzie & Stinson, 1980; Priestley & de Krujff, 1982) have led Crowe & Crowe (1986) to propose a more refined version of Simon's hypothesis. In their view the formation of non-bilayer phases results from the so-called lateral phase separation.

The existence of temperature induced lateral phase separation in membranes is well established (see Quinn, 1985). Membranes are chemically heterogeneous and the fatty acids of particular phospholipids can vary in chain length and degree of saturation. Consequently these different phospholipids have widely different transition temperatures ($Tm$) for the shift which occurs between liquid crystalline and gel phases. As the temperature of a membrane is lowered, lipids with a higher $Tm$ will shift from liquid crystalline to gel phase, lipids with a lower $Tm$ will be excluded from the gel phase until the temperature is lowered to that of their $Tm$. The polar head groups of phospholipids are hydrated and separated from one another by hydrogen-bonded water centred on the phosphate (Hauser et al., 1981). Removal of water will markedly alter the properties of the phospholipids. Crowe & Crowe (1986) propose that dehydration induces lateral phase separations similar to those induced by temperature changes. Thus, as a membrane is dehydrated at constant temperature, lipids having different transition temperatures pass from the liquid crystalline to the gel phase at different water contents and undergo phase separation. This results in the formation of non-bilayer phases and possible displacement of membrane proteins, the consequence being a loss of functional integrity.

Many of the physical changes in membrane structure of cells are reversible and species differences in the degree of disruption of dry membranes may relate to differences in composition, protective mechanisms or to additional damage occurring during desiccation (see below).

### Protein stability and the desiccated state

Much of the work relating to the effect of dehydration on protein stability has been carried out using desiccated tissue, making it difficult to separate direct influences of dehydration from those in which proteins are inactivated through hydrolysis. *In vitro* studies have shown that some proteins undergo extensive losses of activity when freeze-dried. Thus catalase and myosin ATPase were found to lose 87 and 70% of their original activity (Hanafusa, 1969), asparaginase 80% (Marlborough, Miller & Cammack, 1975) and phosphofructokinase 100% (Carpenter, Crowe & Crowe, 1987). Paleg, Stewart & Bradbeer (1984) have shown that glutamine synthetase is precipitated from solution when 'dehydrated' with high molecular weight polyethylene glycol. The photosynthetic enzyme, glyceraldehyde phosphate dehydrogenase was shown to undergo reversible inactivation in mosses subject to dehydration (Stewart & Lee, 1972). Activity was restored by preincubation of extracts with reduced glutathione, suggesting that the loss of activity resulted from the oxidation of essential sulphydryl residues.

Levitt (1962) has of course proposed a theory of frost injury based on the oxidation of sulphydryl groups and the formation of intermolecular disulphide bonds and he has subsequently extended this concept to dehydration injury (see Levitt, 1972). Although popular in the 1960s and 1970s this theory appears to have lost favour in recent years.

### Solute concentrations and desiccation

The increase in solute concentration which occurs during desiccation can be very large and in dehydrated cells major solutes such as potassium may be present at over molar concentrations. Over thirty years ago Lovelock (1953) suggested that lethal injury resulted from the high levels of electrolytes present in dehydrated cells. Solute concentration will have effects on both membrane and protein stability and may be a contributory factor in desiccation injury. However, direct evidence of this is scant. Ionic strength influences the degree of ionisation of the phospholipid head groups and the range of their electrostatic interactions, and an increase of 0.1 M KCl has been shown to alter the transition temperature of dipalmitoylphosphatidylserine by as much as 8 K (MacDonald, Simon &

Baer, 1976). Increasing concentrations of salts have been shown to decrease the stability of some enzymes (see e.g. Ahmad *et al.*, 1982).

## Chemical injury

Over a number of years chemical injury in the form of free radical damage has been invoked as an hypothesis to explain desiccation injury and death. Several biological oxidations, both enzymatic and chemical, generate the free superoxide radical ($O_2^-$) and this can in turn react with hydrogen peroxide to produce singlet oxygen and the hydroxyl radical ($OH^-$), all of which are potential cytotoxic oxidants. The chloroplast is at particular risk from these species because of its high internal oxygen concentration and the presence of molecules which can reduce to $O_2$ to $O_2^-$. Superoxide can be produced enzymically, by enzymes such as xanthine oxidase and galactose oxidase, from the autoxidation of reduced compounds such as flavins, pteridenes, diphenols and ferredoxin. Moreover, isolated thylakoids when illuminated take up oxygen in the absence of added electron acceptors; this 'Mehler' reaction results from the reduction of $O_2$ to $O_2^-$ by electron acceptors of photosystem I (see Asada, Kiso & Toshikawa, 1974). Ferredoxin stimulates this oxygen reduction and in the absence of an $NADP^+$ supply will increase $O_2^-$ formation (see Trebst, 1974). Superoxide is, however, less reactive than other oxygen radicals; in particular, it cannot directly react with membrane lipids. Non-enzymatic dismutation of $O_2^-$ produces oxygen and hydrogen peroxide (Asada *et al.*, 1974). This hydrogen peroxide can in turn react with superoxide to generate the highly reactive hydroxyl radical $OH^-$, particularly when iron salts (Halliwell, 1984) or reduced ferredoxin are present (Elstrer *et al.*, 1978).

Illuminated chlorophyll molecules can form excitation states which are able to transfer energy on to the oxygen molecule, raising it from its ground state to the more reactive excited state known as singlet oxygen.

The production of these reactive species from oxygen has the potential to cause considerable damage to chloroplasts, other membrane systems and cellular macromolecules. Hydroxyl radicals in particular are so reactive they can attack and damage almost every molecule of living cells. Various authors have suggested that peroxidation of unsaturated fatty acids is an underlying cause of membrane damage in desiccation (Simon, 1974; Bewley, 1979; Senaratna & McKenzie, 1986). Peroxidation of pure lipids is very slow but is accelerated in the presence of iron and copper (Sandman & Boger, 1980), haem compounds or illuminated chlorophyll. Peroxidation, once initiated, is autocatalytic and the peroxide and hydroperoxide intermediates decompose with the production of malondialdehyde. Singlet oxygen produced in the chloroplast could react directly with polyunsaturated fatty acid side chains to form lipid peroxides. Free radicals may also bring about de-esterification of phospholipids (Niehaus, 1978). In the hydrated condition there are multiple protective mechanisms against these highly reactive oxygen radicals. These are of three kinds: enzymic, antioxidants and scavengers. There are three enzymes, superoxide dismutase, catalase and peroxidase, which are known to function in preventing the accumulation of oxygen free radicals. Superoxide dismutase removes superoxide with the formation of oxygen and hydrogen peroxide. The latter can be disposed of through the operation of the ascorbate glutathione cycle (see Halliwell, 1984). In the desiccation intolerant moss *Cratoneuron filicinum* catalase and superoxide dismutase decline during dehydration and after rehydration, and this is accompanied by lipid peroxidation and leakage of cellular solutes. The synthesis of antioxidants such as $\alpha$-tocopherol plays a role in protection against free radical induced damage. Tocopherol inhibits lipid peroxidation and is a scavenger of singlet oxygen (see Halliwell, 1984).

Carotenoid pigments can quench the excited states of chlorophyll which result in singlet oxygen formation and they can react with singlet oxygen thus providing a dual protective role. This function is clearly seen in plants treated with herbicides which inhibit carotenoid synthesis (Frosch *et al.*, 1979) and mutant plants lacking carotenoids (Anderson & Robertson, 1960). In both, illumination causes chlorophyll bleaching and severe damage to chloroplast membranes. A variety of low molecular weight compounds can act as free radical scavengers, these include sugars, amino acids and anions. Ascorbate and glutathione are reduced by singlet oxygen (Bodannes & Chan, 1979).

The free radical damage hypothesis of desiccation injury requires that these various protective mechanisms are unable to detoxify reactive species during dehydration and rehydration. There is evidence that free radicals increase with decreasing moisture content of seeds (Priestley *et al.*, 1985), and, in plants subjected to episodic droughting, increased levels of malondialdehyde occur (Price & Hendry, 1987).

*Desiccation tolerance and injury avoidance*

The remarkable tolerance to prolonged anhydrobiosis in resurrection plants suggests they are able to maintain essential structure and physiological integrity in the dry state or are able to repair dehydration-induced damage rapidly following rehydration.

Among resurrection plants two seemingly very different kinds of response occur at the ultrastructural level. In many desiccation tolerant seeds, pollens, mosses and vascular plants, dehydration brings about rather small alterations in mitochondria and chloroplasts and, in general, structural integrity is maintained (see Bewley, 1979). In one species, *Myrothamnus flabellifolia*, unusual ultrastructural features are present (Wellburn & Wellburn, 1976). There is a peculiar arrangement of grana, resembling a staircase, and both chloroplasts and mitochondria have sheath-like structures around them. These might be important in preventing fusion of organelle membranes in the dehydrated state.

In contrast, there are others such as *Borya nitida* (Gaff, Zee & O'Brien, 1976) and *Xerophyta villosa* (Hallam & Luff, 1980) which undergo very extensive loss in ultrastructural integrity. The bounding membranes or organelles become ill defined, and numbers of thylakoids and cristae are reduced. The nuclear membrane and tonoplast are preserved but the vacuole is fragmented. As mentioned earlier, failure to undergo this apparent 'damage' results in a loss of the plant's extreme desiccation tolerance. The restoration or repair of ultrastructural organisation is rapid. Within two hours, organelle bounding membranes have a more distinct appearance, by eight hours mitochondrial cristae are visible and between 24 and 48 h chloroplast thylakoids are again present (Hallam & Gaff, 1978). Reorganisation of cell structure on rehydration appears to involve rehabilitation or repair of degraded organelles rather than the production of new organelles. Evidently some species can restore not only the ultrastructural changes induced by desiccation but also they can rapidly resynthesise chlorophyll and other pigments lost during dehydration.

These extensive alterations in cell structure and the biochemical machinery are indicative of entry into an 'ametabolic' condition. In this condition damage from free radicals is potentially decreased, certainly the loss of chlorophyll and chloroplast structure removes a major source of free radical generation. About 50% of the extremely desiccation tolerant monocots exhibit extensive loss of chlorophyll and ultrastructural organisation when desiccated. Dicots, ferns and bryophytes retain most of their chlorophyll and exhibit small changes in structure when dry (see Gaff, 1977).

Evidence that species differences in membrane composition or protein structure contribute to the tolerance of resurrection plants is generally lacking. Membrane restructuring of the kind reported for temperature adaptation (see Storey & Storey, 1988) does not appear to have been reported for desiccation tolerant plants. Presumably the suspension of metabolic activity in the anhydrobiotic state mitigates against selection of novel membrane configurations. However, stabilisation of membranes and proteins through the synthesis and accumulation of low molecular weight 'protectants' is thought by some to be of critical importance in desiccation tolerance. Following the work of Warner (1962) and Webb (1965), Crowe (1971) put forward the so-called water replacement hypothesis in which it was suggested that polyhydroxyl compounds replaced the structural water of cellular components. The experimental observations underlying this hypothesis were that many animals, able to undergo reversible drying, contained high concentrations of compounds such as glycerol, trehalose and sugar alcohols (see Crowe & Clegg, 1973, 1978). In recent years much interest has focused on the role of trehalose which can constitute more than 20% of the dry weight of some desiccation resistant animals (Madin & Crowe, 1975). Trehalose is reported to inhibit the fusion of vesicles which can occur during drying, maintain lipids in a fluid phase and stabilise labile proteins (see Crowe *et al.*, 1987). Carpenter, Crowe & Crowe (1987) have shown that increasing concentrations of trehalose, maltose and sucrose are effective in promoting the recovery of phosphofructokinase activity after freeze-drying of the enzyme. In this study considerable synergism was observed between sugars and transition metals such as zinc. Indeed, the concentration of sugars giving maximum protection was much lower when 0.9 mM $Zn^{2+}$ was present and the metal was essential for galactose and glucose to exhibit any protective effect. Comparable protective effects have been reported for other solutes that can accumulate in stressed plants. Proline, for example, has been shown to ameliorate the deleterious effects of heat, pH, salt, chemicals and dehydration on enzyme activity and organelle systems (Ahmad *et al.*, 1982; Bowlus & Somero, 1979; Nash, Paleg & Wiskich, 1982; Paleg *et al.*, 1981; Paleg *et al.*, 1984; Yancey & Somero, 1979).

The mechanism by which these solutes exert their influence on protein stability is uncertain. The phenomenon has been extensively studied by Timasheff and his colleagues and their conclusion is that all of the protein structure-stabilising compounds are preferentially excluded from contact with the surface of the protein (Timasheff, 1982). This explanation is rather different from that invoked in the water replacement hypothesis.

Trehalose and other low molecular weight compounds have been shown to stabilise membrane and bilayer systems against dehydration-induced damage. Unilamellar phospholipid vesicles which have been dehydrated leak their contents when rehydrated. In the presence of trehalose, sucrose, proline and glycine betaine leakage is prevented (Rudolph, Crowe & Crowe, 1986). Again the molecular mechanisms underlying these effects are not completely understood. Trehalose inhibits phase transitions associated with dehydration of lipid bilayer (Rudolph *et al.*, 1986) and this may be linked to the ability of trehalose to form hydrogen bonds with the polar head groups of phospholipids. Proline action is suggested to involve effects on the hydration layer surrounding phospholipids and possibly also its intercalation between phospholipid head groups (Rudolph *et al.*, 1986).

Although trehalose is conspicuously absent in most vascular plants and bryophytes (Lewis, 1984), a variety of environmental stresses induce the accumulation of compounds which *in vitro* at least exercise a protective role (see Stewart & Larher, 1981). High concentrations of proline are characteristic of many water- or salt-stressed plants. In addition, high concentrations of proline are common in the pollen of many species (Stewart & Larher, 1981). However, there is no evidence that proline accumulates to high levels in resurrection plants (Tymms & Gaff, 1979). Glycine betaine, like proline, is of widespread occurrence, accumulates in stressed tissues and has protective properties (Wyn Jones & Storey, 1981; Paleg *et al.*, 1981, 1984). It does not appear to have been reported in desiccation tolerant plants. Several low molecular weight carbohydrates accumulate in water- or salt-stressed tissue and large concentrations of sorbitol (Ahmad, Larher & Stewart, 1979), mannitol and pinitol (Popp, 1984) have been reported in some plants.

Polyols are present in desiccation tolerant lichens and liverworts, although not in mosses (Lewis, 1984). More generally starch hydrolysis and sugar accumulation occur in many plants experiencing severe water deficits (Hsiao, 1973). It is tempting to speculate that the accumulation of low molecular weight solutes in reponse to water stress represents a mechanism for the protection of membranes and proteins in the dry state.

It is striking that many of the solutes which accumulate in stressed plants and which have protective properties are also reported to reduce free radical activity. Implicit in the theory of free radical induced damage under conditions of dehydration is the notion that the cell's defensive mechanisms break down or are overloaded. If this theory is correct then it follows that organisms able to tolerate extreme desiccation are able to maintain their defensive mechanisms or amplify them under conditions of increased stress. There is some evidence that there may be increased synthesis of enzymes which detoxify free radicals. During slow desiccation of the dehydration-tolerant moss, *Tortula ruralis*, superoxide dismutase and catalase activities increase whereas in the intolerant species, *Cratoneuron filicinum*, these two activities decline (Bewley, 1979). Recently Price & Hendry (1987) have suggested that $\alpha$-tocopherol synthesis is stimulated in water-stressed plants. They were unable to demonstrate consistent changes in the activities of superoxide dismutase, peroxidase or catalase and concluded that enhanced tocopherol levels gave protection against oxidative damage to membranes.

Proline is also able to detoxify free radicals by forming long-lived adducts with them (Floyd & Zs-Nagy, 1984). Another group of compounds with radical scavenging properties are the polyamines (Drolet *et al.*, 1986). Superoxide radical formation from chemical and enzyme reactions is inhibited by spermine, spermidine, putrescine and cadavarine when these are present at 10–50 mM. Moreover, Kitada *et al.* (1979) have reported that spermine inhibits lipid peroxidation in microsomes. The accumulation of di- and polyamines in response to environmental stress including water deficits is well documented (Smith, 1985).

### Concluding remarks

Desiccation tolerant species appear to fall into two categories, those which retain in large part their ultrastructural organisation during drying and those which undergo extensive losses of chlorophyll and cellular organisation. Membrane and metabolic disfunction which occurs in the transition from the dry state to full hydration is ameliorated by the production of protectant molecules which interact with membranes and proteins to preserve their integrity during rapid rehydration. Oxidative damage to membranes and proteins as the result of free radical attack is reduced by increased free radical scavengers. Some of the compounds which have stabilising effects on membranes and macromolecules may also act as free radical scavengers.

Among the wider implications of desiccation tolerance are the preservation of genetic resources. In recent years agriculturalists and basic scientists have recognised the importance of preserving plant germplasm. Withers (1980) has pointed out that the greatest risks associated with preservation are that physiological deterioration may occur in storage and that environmental stress may depress the subsequent performance of preserved material. Research into desiccation tolerance is an important aspect of preservation, particularly since the survival of material at sub-zero and liquid nitrogen temperatures ($-196$ K) is dependent on its water content (Stanwood, 1986). Particularly relevant in the present context are results showing enhanced survival of seeds stored in liquid nitrogen after being dehydrated with high concentrations of proline (3–5 M; Stanwood, 1986). An understanding of desiccation resistance in resurrection plants may assist in improving preservation methods used for the conservation of plant genetic resources.

### References

Ahmad, I., Larher, F. & Stewart, G.R. (1979). Sorbitol A compatible osmotic solute in *Plantago maritima*. *New Phytologist*, **82**, 671–8.

Ahmad, I., Larher, F., Mann, A.F., McNally, S.F. & Stewart, G.R. (1982). Nitrogen metabolism of halophytes. IV. Characteristics of glutamine synthetase from *Triglochin maritima* L. *New Phytologist*, **91**, 585–95.

Anderson, I.C. & Robertson, D.S. (1960). Role of carotenoids in protecting chlorophyll from photodestruction. *Plant Physiology*, **35**, 531–4.

Asada, K., Kiso, K. & Toshikawa, K. (1974). Univalent reduction of molecular oxygen by spinach chloroplasts on illumination. *Journal of Biological Chemistry*, **249**, 2175–81.

Bewley, J.A. (1979). Physiological aspects of desiccation tolerance. *Annual Review of Plant Physiology*, **30**, 195–238.

Binder, W.D., Mitchell, G.M. & Ballantyne, D.J. (1974). Pollen viability testing, storage and related physiology. Canada Forestry Service, Pacific Forest Research Centre, Victoria, B.C. Report BC-X-105, pp. 1–37.

Bodannes, R.S. & Chan, P.C. (1979). Ascorbic acid as a scavenger of singlet oxygen. *FEBS Letters*, **105**, 195–6.

Bowlus, R.D. & Somero, G.N. (1979). Solute compatibility with enzyme function and structure: rationales for the selection of osmotic agents and end products of anaerobic metabolism in marine invertebrates. *Journal of Experimental Zoology*, **208**, 137–52.

Bradbury, J.H., Foster, J.R., Hammer, B., Lindsay, J. & Murrell, W.G. (1981). The source of heat resistance of bacterial spores. Study of water in spores by NMR. *Biochimica Biophysica Acta*, **678**, 157–64.

Carpenter, J.F., Crowe, L.M. & Crowe, J.H. (1987). Stabilization of phosphofructokinase with sugars during freeze drying: characterization of enhanced protection in the presence of divalent cations. *Biochimica Biophysica Acta*, **923**, 109–15.

Crittenden, P.D. & Kershaw, K.A. (1978). A procedure for the simultaneous measurement of net $CO_2$-exchange and nitrogenase activity in lichens. *New Phytologist*, **80**, 393–401.

Crowe, J.H. (1971). Anhydrobiosis: an unsolved problem. *American Naturalist*, **105**, 563–73.

Crowe, J.H. & Clegg, J.S. (ed.) (1973). *Anhydrobiosis*. Stroudsburg, P.A.: Dowden Hutchinson & Ross.

Crowe, J.H. & Clegg, J.S. (ed.) (1978). *Dry Biological Systems*. New York: Academic Press.

Crowe, L.M. & Crowe, J.H. (1986). Hydration-dependent phase transitions and permeability properties of biological membranes. In *Membranes, Metabolism and Dry Organisms*, ed. A.C. Leopold, pp. 210–30. Ithaca, N.Y.: Cornstock Publishing Associates.

Crowe, J.H., Crowe, L.M., Carpenter, J.F. & Aurell Wistrom, C. (1987). Stabilization of dry phospholipid bilayers and proteins by sugars. *Biochemical Journal*, **242**, 1–10.

Dilks, T.J.K. & Proctor, M.C.F. (1974). The pattern of recovery of bryophytes after desiccation. *Journal of Bryology*, **8**, 97–115.

Drolet, G., Dumbroff, E.B., Legge, R.L. & Thompson, J.E. (1986). Radical scavenging properties of polyamines. *Phytochemistry*, **25**, 367–71.

Elstrer, E.F., Saran, H., Bors, W. & Longfelder, E. (1978). Oxygen activation in isolated choroplasts. *European Journal of Biochemistry*, **89**, 61–6.

Farrer, J.F. (1976). Ecological physiology of the lichen *Hypogymnia phipodes*. 11. Effects of wetting and drying cycles and the concept of 'physiological buffering'. *New Phytologist*, **77**, 105–13.

Floyd, R.A. & Zs-Nagy (1984). Formation of long lived hydroxyl free radical adducts of proline and hydroxyproline in a Fenton reaction. *Biochimica Biophysica Acta*, **790**, 94–7.

Frosch, S., Jabben, M., Bergfeld, R., Klanig, H. & Mohr, A. (1979). Inhibition of carotenoid biosynthesis by the herbidice SAN 9789 and its consequences for the action of phytochrome on plastogenesis. *Planta*, **145**, 497–505.

Gaff, D.F. (1971). The desiccation tolerant higher plants of southern Africa. *Science*, **174**, 1033–4.

Gaff, D.F. (1977). Desiccation tolerant vascular plants of southern Africa. *Oecologia*, **31**, 95–109.

Gaff, D.F. & Churchill, D.M. (1976). *Borya nitida* Labill. – an Australian species in the Liliaceae with desiccation-tolerant leaves. *Australian Journal of Botany*, **24**, 209–24.

Gaff, D.F. & Latz, P. (1978). The occurrence of resurrection plants in the Australian flora. *Australian Journal of Botany*, **26**, 483–92.

Gaff, D.F., Zee, S.T. & O'Brien, T.P. (1976). The fine structure of dehydrated and reviving leaves of *Borya nitida* Labill. – a desiccation tolerant plant. *Australian Journal of Botany*, **24**, 225–36.

Goss, J.A. (1968). Development, physiology and biochemistry of corn and wheat pollen. *Botanical Reviews*, **34**, 333–58.

Hallam, N.D. & Gaff, D.F. (1978). Reorganization of fine structure during rehydration of desiccated leaves of *Xerophyta villosa*. *New Phytologist*, **81**, 349–55.

Hallam, N.D. & Luff, S.E. (1980). Fine structure changes in mesophyll tissue of the leaves of *Xerophyta villosa* during desiccation. *Botanical Gazette*, **141**, 173–9.

Halliwell, B. (1984). *Chloroplast Metabolism*. Oxford: Clarendon Press.

Hanafusa, N. (1969). Denaturation of enzyme proteins by freeze drying. In *Freezing and Drying of Microorganisms*, ed. T. Nei, pp. 117–29. Baltimore, MD: University Park Press.

Hauser, H., Pascher, I., Pearson, R.H. & Sundell, S. (1981). Preferred conformation and molecular packing of phosphatidyl ethanolamine and phosphatidylcholine. *Biochimica Biophysica Acta*, **650**, 21–51.

Hsiao, T.C. (1973). Plant responses to water stress. *Annual Review of Plant Physiology*, **24**, 519–70.

Iljin, W.S. (1931). Austrobaung resisterz des Farnes Notochlaena Morantae R. Br. *Protoplasma*, **13**, 352–66. In *Strategies of Microbial Life in Extreme Environments*, ed. M. Shito, pp. 163–77. Weinheim: Verlag-Chemie.

Keijzer, C.J. (1983). Hydration changes during anther development. In *Pollen: Biology and Implications for Plant Breeding*, ed. N.L. Matching & F.A. Ohaviane, pp. 199–202. Elsevier.

Kershaw, K.A. (1972). The relationship between moisture content and net assimilation rate of lichen thalli and its ecological significance. *Canadian Journal of Botany*, **50**, 543–55.

Kitada, M., Igarashi, K., Hirose, S. & Kitagawa, H. (1979). Inhibition by polyamines of lipid peroxide formation in rat liver microsomes. *Biochemical & Biophysical Research Communications*, **87**, 388–92.

Lange, O.L. (1969). Ecophysiological investigation of lichens of the Negev desert. I. $CO_2$ gas exchange of *Ramalina maciformis* (Del.) Borg under controlled conditions of the laboratory. *Flora*, **158**, 324–59.

Lange, O.L., Schutze, E.D., Kappen, L., Buschborn, U. & Evenori, M. (1975). Adaptations of desert lichens to drought and extreme temperatures. In *Environmental Physiology of Desert Organisms*, ed. N.F. Hadley, pp. 20–37. Strandsberg, P.A.: Dowden Hutchison & Ross.

Levitt, J. (1962). A sulfhydryl-disulfide hypothesis of frost injury and resistance in plants. *Journal of Theoretical Biology*, 3, 355–91.

Levitt, J. (1972). *Responses of Plants to Environmental Stresses*. New York: Academic Press.

Lewis, D.H. (1984). Occurrence and distribution of storage carbohydrates in vascular plants. In *Storage Carbohydrates in Vascular Plants*, ed. D.H. Lewis, pp. 1–32. Cambridge: Cambridge University Press.

Lovelock, J.E. (1953). The mechanism of the protective section of glycerol against haemolysis by freezing and thawing. *Biochemica Biophysica Acta*, 11, 28–36.

MacDonald, R.C., Simon, S.P. & Baer, E. (1976). Ionic influences on the phase transitions of dipalmitoyl phosphatidylserine. *Biochemistry*, 15, 885–97.

Madin, K.A.C. & Crowe, J.H. (1975). Anhydrobiosis in nematodes: carbohydrate and lipid metabolism during dehydration. *Journal of Experimental Zoology*, 193, 335–41.

Marlborough, D.I., Miller, D.S. & Cammack, K.A. (1975). Comparative study on conformational stability and sub-unit interactions of two bacterial asparaginases. *Biochimica Biophysica Acta*, 386, 576–89.

McKenzie, B.D. & Stinson, R.H. (1980). Effect of dehydration on leakage and membrane structure in *Lotus corniculatus* L. seeds. *Plant Physiology*, 66, 316–21.

Nash, D., Paleg, L.G. & Wiskich, J.T. (1982). The effect of proline, betaine and some other solutes on the heat stability of mitochondrial enzymes. *Australian Journal of Plant Physiology*, 9, 45–57.

Niehaus, W.J., Jr (1978). A proposed role of superoxide anion as a biological nucleophile in the deesterification of phospholipid. *Bioorganic Chemistry*, 7, 77–84.

Paleg, L.G., Douglas, T.J., Van Daal, A. & Keech, D.B. (1981). Proline and betaine protect enzymes against heat inactivation. *Australian Journal of Plant Physiology*, 8, 107–14.

Paleg, L.G., Stewart, G.R. & Bradbeer, J.W. (1984). Proline and glycine betaine influence protein solvation. *Plant Physiology*, 75, 974–8.

Popp, M. (1984). Chemical composition of Australian mangroves. II. low molecular weight carbohydrates. *Zeitschrift für Pflanzenphysiologie*, 113, 411–21.

Price, A. & Hendry, G. (1987). The significance of the tocopherols in stress survival in plants. In *Free Radicals, Oxidant Stress and Drug Action*, ed. C. Rice Evans, pp. 443–50. London: Richelieu Press.

Priestley, D.A. & de Krujff, B. (1982). Phospholipid notional characteristics in a dry biological system. *Plant Physiology*, 70, 1075–9.

Priestley, D.A. & Posthumus, M.A. (1982). Extreme longevity of *Lotus* seeds from Palantien. *Nature*, 299, 148–9.

Priestley, D.A., Warner, B.G., Leopold, A.C. & McBride, M.B. (1985). Organic free radical levels in seeds and pollen. The effects of hydration and ageing. *Physiologia Plantarum*, 70, 88–94.

Proctor, M.C.F. (1982). Physiological ecology: water relations, light and temperature responses, carbon balance. In *Bryophyte Ecology*, ed. A.J.E. Smith, pp. 333–81. London: Chapman & Hall.

Quinn, P.J. (1985). A lipid-phase separation model of low temperature damage to biological membranes. *Cryobiology*, **22**, 128–40.

Roberts, E.H. (1973). Predicting the storage life of seeds. *Seed Science & Technology*, **1**, 499–514.

Ross, K.F.A. & Billing, E. (1957). The water and solid content of living bacterial spores and vegetative cells as indicated by refractive index measurements. *Journal of General Microbiology*, **17**, 418–25.

Rudolph, A.S., Crowe, J.H. & Crowe, L.M. (1986). Effects of three stabilizing agents – proline, betaine and trehalose on membrane phospholipids. *Archives of Biochemistry and Biophysics*, **245**, 134–43.

Sandman, G. & Boger, P. (1980). Copper mediated lipid peroxidation processes in photosynthetic membranes. *Plant Physiology*, **66**, 797–800.

Scott, G.A.M. (1982). Desert Bryophytes. In *Bryophyte Ecology*, ed. A.J.E. Smith, pp. 105–22. London: Chapman & Hall.

Senaratna, T. & McKenzie, B.D. (1986). Loss of desiccation tolerance during seed germination: A free radical mechanism of injury. In *Membranes, Metabolism and Dry Organisms*, ed. A.C. Leopold, pp. 85–101. Ithaca, N.Y.: Cornstock Publishing Associates.

Simon, E.W. (1974). Phospholipids and plant membrane permeability. *New Phytologist*, **73**, 377–420.

Smith, T.A. (1985). Polyamines. *Annual Review of Plant Physiology*, **36**, 117–43.

Stanwood, P.C. (1986). Dehydration problems associated with the preservation of seed and plant germplasm. In *Membranes, Metabolism and Dry Organisms*, ed. A.C. Leopold, pp. 327–40. Ithaca, N.Y.: Cornstock Publishing Associates.

Steinbrick, C. (1900). Zur terminologie der volum-anderungen pflanzlicher Gewebe und organischer substanzen bei wechselndem Fluosigkeitsgehalt. *Ber. Deut. Bot. Ges.*, **18**, 217–24.

Stewart, G.R. & Larher, F. (1981). The accumulation of amino acids and related compounds in relation to environmental stress. In *Biochemistry of Plants, Vol. V*, ed. B.J. Miflin, pp. 609–35. London: Academic Press.

Stewart, G.R. & Lee, J.A. (1972). Desiccation injury in mosses. II. The effect of moisture stress on enzyme levels. *New Phytologist*, **71**, 461–6.

Storey, K.B. & Storey, J.M. (1988). Freeze tolerance in animals. *Physiological Reviews*, **68**, 27–84.

Timasheff, S.N. (1982). Preferential interactions in protein–water-cosolvent systems. In *Biophysics of Water*, ed. F. Franks, pp. 70–2. London: John Wiley.

Trebst, A. (1974). Energy conservation in photosynthetic electron transport of chloroplasts. *Annual Review of Plant Physiology*, **25**, 423–58.

Tymms, M.J. & Gaff, D.F. (1979). Proline accumulation during water stress in resurrection plants. *Journal of Experimental Botany*, **30**, 165–8.

Warner, D.T. (1962). Some possible relationships of carbohydrates and other biological compounds with water structure at 37°. *Nature*, **196**, 1053–8.

Webb, S.J. (1965). Bound water. In *Biological Integrity*, ed. C.C. Thomas, pp. 215–30. Springfield, Illinois: Charles C. Thomas.

Wellburn, F.A.M. & Wellburn, A.R. (1976). Novel chloroplasts and unusual cellular ultrastructure in the 'resurrection' plant *Myrothamnus flabellifolia* Welw. (Myrothamnaceae). *Botanical Journal of the Linnean Society*, **72**, 51–4.

Withers, L.A. (1980). Tissue culture storage for genetic conservation, IBPGR Report. International Bureau of Plant Genetic Resources, Food and Agriculture Organization, Rome, Italy, 91 pp.

Wyn Jones, R.G. & Storey, R.H. (1981). Betaines. In *The Physiology and Biochemistry of Drought Resistance in Plants*, ed. L.G. Paleg and D. Aspinall, pp. 171–204. London: Academic Press.

Yancey, R.N. & Somero, G.N. (1979). Counteraction of urea destabilization of protein structure by methylamine osmoregulatory compounds of elasmobranch fishes. *Biochemical Journal*, **183**, 317–23.

STEPHEN G. HUGHES, JOHN A. BRYANT
AND NICHOLAS SMIRNOFF

# 8 Molecular biology: application to studies of stress tolerance

## Introduction

Research on genetic manipulation of plants has accelerated drama-
tically during the past five years. This has exploited 'natural' gene delivery
systems, leading to the development of vectors, and a number of 'vectorless'
DNA delivery systems. Further, since plants show extensive powers of
regeneration it is, for many species, possible to grow whole plants from
genetically engineered cells. Formation of flowers on the genetically
engineered plants, followed by pollination, fertilisation, embryogenesis and
seed set will then lead to the transmission of the acquired gene to the next
generation, provided of course that the gene in question is stably incorporated into the host plant's genome.

In this chapter we first discuss the development of systems for delivering
DNA to plant cells. The remainder of the chapter then outlines the ways in
which molecular genetics can be applied to research on stress tolerance,
giving examples where these techniques are being used.

### Systems for gene transfer

For a detailed discussion of DNA delivery systems, readers are
referred to the recent very comprehensive review by Walden (1988). Here
we concentrate on the two methods most used in stress tolerance studies.

Agrobacterium tumefaciens *and its tumour-inducing plasmid*

*Agrobacterium tumefaciens* is a soil-dwelling bacterium which
readily infects a wide range of dicotyledonous plants, usually gaining entry
via a wound site. Infection causes the growth of a tumour, usually at the
original infection site (Fig. 1) and once the tumour is established it will grow
in the absence of bacterial cells. This means that the originally infected plant
cells have been induced to divide and that the cells arising from this division
have, in some way, been transformed so as to evade the host plant's normal
control mechanisms. A further indication that some sort of transformation
occurs is that the tumour cells synthesise unusual amino acid derivatives,

called opines, the particular opine produced being related to the particular strain of *Agrobacterium*. The plant host is unable to metabolise opines, but *Agrobacterium* can use these compounds as carbon and nitrogen sources. Thus, the infection has resulted in the formation of a large number of cells which produce nutrients for the bacterium (Fig. 1).

Fig. 1. Infection and transformation of a suitable host plant by *Agrobacterium tumefaciens* (from Bryant, 1988).

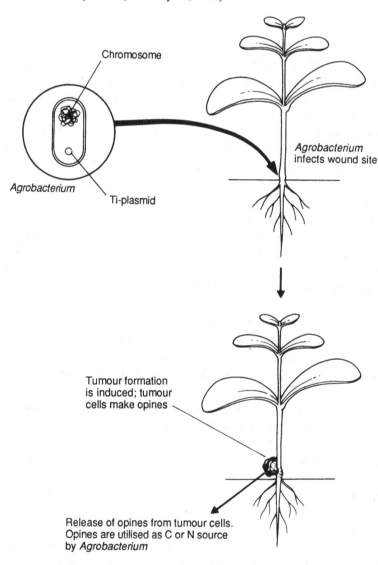

This transformation of the host cells is based on the presence in *A. tumefaciens* of a plasmid, the tumour-inducing or Ti-plasmid. This is much larger than the types of plasmid vector used in genetic manipulation of *E. coli*. The latter are generally in the range 2–8 kbp (1 kbp = 1000 base pairs) whereas the Ti plasmid has a size of *c*. 200 kbp. From the point of view of genetic transformation, two regions of the plasmid are important. The first of these is the T-region, consisting of *c*. 20 kbp of DNA bounded by 25 bp repeats at each end (Fig. 2). Following a successful infection by *Agrobacterium*, this region is excised from the plasmid and becomes integrated into the host's nuclear DNA. The T-DNA carries a number of

Fig. 2. Schematic representation of the Ti-plasmid of *Agrobacterium*, showing the genes involved in transformation of host plants. Note that the genetic map is not to scale: in reality, the T-region makes up about 10% of the total plasmid genome (from Bryant, 1988).

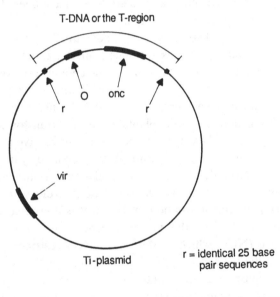

T-DNA or the T-region

Ti-plasmid

r = identical 25 base
    pair sequences

O = gene coding for opine-
    synthesising enzyme

onc = group of genes that induce
    uncontrolled division (tumour
    formation) in the
    host plant

vir = group of genes that control
    the transfer of the T-DNA
    to the host chromosome

genes, including one gene which codes for an opine-synthesising enzyme (e.g. nopaline synthase) and three, or in some strains, four, *onc* genes. The latter code for enzymes which are involved in biosynthesis of plant hormones and hence confer on the cells the ability to divide in an uncontrolled manner. Another region of the plasmid, the *vir* region, although not transferred to the host's genome, is essential for the transfer process.

The Ti-plasmid is thus a natural vector for genetic engineering of dicotyledonous plants. However, there are two very obvious problems. The first is that the size of the plasmid, 200 kbp, makes it very difficult to manipulate, not least because of the very large number of restriction enzyme sites present in such a large plasmid. The second is that the transformed phenotype is aggressively tumorous, and it is therefore very difficult to manipulate the growth conditions so as to regenerate whole plants from the transformed cells. Several strategies have been developed to overcome these problems (Schell, 1987).

The first and most general strategy is to remove the T-DNA region and insert it into a normal laboratory plasmid, such as pBR322. In this form, the T-DNA can be cloned in *E. coli*. This will provide enough T-DNA to work with. Next, it is usual to excise the *onc* genes. This will ensure that if host plant cells are transformed with recombinant T-DNA, then it will be more straightforward to generate whole plants from the transformed cells. Thirdly, the remaining T-DNA needs to be tailored to receive the foreign genes. It should be noted here that although the nopaline synthase (*nos*) gene is in a bacterial plasmid, it has promoter and termination signals which are recognised by the plant host. The *nos* gene can therefore be used as a marker or reporter for transformation, although it is non-selectable. If a selectable marker is needed, the bacterial gene coding for neomycin phosphotransferase (NPT, which confers kanamycin resistance) is often spliced in between the *nos* promoter and termination signals, after excision of the *nos* coding region, thus bringing the bacterial gene under the control of typical plant-type regulatory sequences. In fact, any gene can be spliced into this position but for many purposes it may be necessary to have the inserted gene under the control of its own regulatory sequences. To this end, many versions of the T-DNA have been made with a 'polylinker' site situated away from the *nos* gene. The polylinker site is constructed so as to provide recognition signals for several restriction enzymes over a very short piece of DNA, thus providing a 'universal' cloning site.

Finally, it must be emphasised that a plasmid such as pBR322 carrying a 'tailored' T-DNA region does not contain the *vir* genes which are essential for the successful transfer of the T-DNA into the host's genome. This problem is overcome by mobilising the tailored vector back into *Agrobacterium*, which carries a Ti-plasmid. If the Ti-plasmid is complete, then the T-DNA-derived sequence in pBR322 has a high chance of reinserting into the Ti-plasmid via homologous recombination at the repeated border sequences. This system is known as the co-integrative system. Alternatively, the tailored pBR322-T-DNA plasmid may be mobilised into *Agrobacterium* cells which carry a Ti-plasmid from which the T-DNA has been removed. The *vir* region on this plasmid is still effective in promoting transfer of the T-DNA to the host's genome, even though the T-DNA is now in a different molecule. Systems such as these are known as binary vector systems. Both these systems have been used successfully to transfer foreign genes to plant cells.

The genetic engineering of plants with recombinant T-DNA is generally

Fig. 3. One of the experimental systems in common use for genetic engineering of plants with *Agrobacterium* containing a recombinant Ti-plasmid (from Bryant, 1988).

Leaf discs are floated on a suspension of *Agrobacterium* cells that contain genetically engineered Ti-plasmids. The bacterium infects cells around the edges of the discs, and the T-DNA is transferred to the chromosomes of the infected cells.

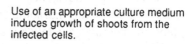

Use of an appropriate culture medium induces growth of shoots from the infected cells.

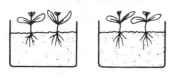

The shoots are transferred to a root-inducing medium, leading to the formation of plantlets.

The plantlets are grown up to give genetically engineered mature plants.

achieved by floating leaf discs on a suspension of *Agrobacterium* carrying the appropriate plasmid. The *Agrobacterium* invades the wounded edge of the discs and, after a limited amount of cell division, shoots may be induced to grow out from the transformed cells at the edges. The shoots may then be induced to root by provision of an appropriate culture medium. Plants which are genetically engineered in this way, with genes placed under the control of the *nos* promoter, or of other normal plant promoters, show general expression of the genes throughout the plant. The transferred genes are integrated into the host's nuclear DNA, and are inherited according to normal Mendelian genetics (Fig. 3).

Another useful experimental system is likely to develop from a very recent series of experiments by Roberta Smith's group at Texas A & M University (Ulian *et al.*, 1988). In these experiments, isolated shoot apices from *Petunia* seedlings were transformed by infection with *Agrobacterium* harbouring a Ti-plasmid carrying genes for kanamycin resistance and β-glucuronidase. Regeneration was rapid, and the transgenic plants expressed the foreign genes in all tissues. As pointed out by the authors, this methodology will be particularly useful for those plants which are difficult to regenerate from tissues other than meristems, the latter being the tissues of choice for 'difficult' species.

For some purposes, it may be necessary to get the inserted gene to work in particular parts of the plant, or at particular stages of the plant's life or, as in stress tolerance, in response to particular environmental conditions. It is in connection with this that some of the most exciting developments in plant genetic engineering have taken place. One example will suffice to illustrate this. The enzyme ribulose 1,5 bisphosphate carboxylase (Rubisco) is located in the chloroplast where it acts as the primary carboxylating enzyme in photosynthesis. The enzyme is composed of two types of protein subunit: large (*c.* 50–58 kDa), which are encoded in chloroplast DNA and synthesised in the chloroplast, and small (*c.* 12–18 kDa), which are encoded in the nucleus, synthesised in the cytoplasm and transported into the chloroplast. The genes coding for the small subunit (*ssu*) are expressed only in light, and only in green cells. This light-dependent expression is under the control of the plant light receptor molecule, phytochrome.

The regulation of the expression of the small subunit gene by light is the basis of achieving the selective expression of 'foreign' genes in green cells in a light-dependent manner. 'Upstream' of the promoter region of the small subunit gene is a further gene control region, extending back to about 975 bases before the start of the gene. This region is the key to light-dependent, green-cell specific expression of the *ssu* gene (Herrera-Estrella *et al.*, 1985). A foreign gene inserted into T-DNA adjacent to the *ssu* promoter and then transferred to a host plant is expressed throughout the plant. If, however, the recombinant gene construction also contains the 975 bp upstream flanking region, then the foreign gene is expressed in the genetically engineered plants only in a light-dependent green-cell specific manner. Control sequences of the type represented by the upstream flanking region of the *ssu* gene are known as enhancers. Their role in plants was, in fact, discovered by genetic engineering experiments, and enhancers which are specific for seeds, for storage organs and for flowers are also known.

Enhancers then may well prove to be the key to obtaining expression of foreign genes in the right cells at the right time in genetically engineered plants. However, further levels of sophistication are possible. It was noted above that the Rubisco small subunit is transported from the cytoplasm to the chloroplast. This happens as follows: the nascent protein, synthesised on cytoplasmic ribosomes, is larger than the final protein product. The extra length is caused by the presence of *c*. 15 extra amino acids at the N-terminal end. These extra amino acids constitute a transit signal peptide which interacts specifically with the chloroplast envelope, facilitating transport of the protein into the chloroplast; during the passage across the envelope, the transit peptide is removed. These findings are the basis of experiments in which a foreign gene is spliced to the region of the *ssu* gene coding for the transit peptide leading to the transfer of the foreign protein into the chloroplast (Herrera-Estrella *et al.*, 1985; Van den Broek *et al.*, 1985). Similar transit peptides are known for proteins which are transported to other cellular organelles, such as mitochondria. This means that not only is it possible to get a foreign gene switched on in the right part of the plant at the right time, but it is also possible, if necessary, to target the gene product to a specific subcellular organelle.

Use of *Agrobacterium* carrying recombinant plasmids with engineered T-DNA is currently one of the most promising methods for the commercial application of plant genetic engineering. The wide host range of *Agrobacterium* makes this a suitable system for use with many dicotyledonous plants. To date, most experiments have been performed with plants which are very easy to regenerate, particularly members of the Solanaceae, such as potato, tomato, tobacco and *Petunia*, plus a few other species including sunflower, but there is clearly potential to apply these techniques to many other broadleaved crops such as soybean, rape and even tree species.

Under natural conditions, *Agrobacterium* does not appear to infect monocotyledonous plants. Genetic manipulation of cereal crops using a Ti-plasmid delivered by *Agrobacterium* therefore seems out of the question. However, there are some data to suggest that *Agrobacterium* can, under certain circumstances, infect monocots, but that it does not cause crown gall formation; on superficial examination, the plant will appear uninfected. However, direct injection of *Agrobacterium* into tissues of *Chlorophytum*, *Narcissus* and *Asparagus* does lead to low-level production of opines by the plant tissues, suggesting that some form of transformation takes place (Hooykaas-van Slogteren, Hooykaas & Schilperoort, 1984; Hernalsteens *et al.*, 1985). More interesting still is the observation that if two adjacent double-stranded copies of the genome of maize streak virus (maize streak virus, MSV, a member of the gemini virus group, is normally single-stranded) are spliced into the T-DNA followed by infection of maize seedlings with *Agrobacterium* harbouring the Ti-MSV plasmid, then the maize seedlings develop all the symptoms of an MSV infection (Grimsley *et al.*, 1987). This has become known as *Agro-infection* and is all the more intriguing in view of the fact that, normally, MSV is obligately transmitted by leaf hoppers.

Although it is now becoming clear that *Agrobacterium* can, under some circumstances, infect monocot plants and deliver DNA to them, it is not known whether the delivered DNA integrates into the host plant's genome. MSV DNA, for example, does not, in normal infection cycles, integrate, and the expression of symptoms seen after agro-infection can thus occur without integration. What all this adds up to is that there is still a very long way to go before the *Agrobacterium*–Ti-plasmid system can be used as a vector for monocots.

### 'Vectorless' DNA transfer

An early feature to emerge from work on genetic manipulation of animal cells was that many types of animal cell can, after appropriate treatment, take up DNA without the use of a vector, with the DNA being subsequently integrated into the genome of the recipient cells (Watson, Tooze & Kurtz, 1983). This has led to similar approaches being used for plant cells. For plant cells, it is first necessary to remove the cell wall with enzymes to produce protoplasts. Then, since plant cells, unlike animal cells, do not take up DNA precipitated onto their surface by calcium phosphate, it is necessary to make the protoplast membranes leaky. Initial experiments used polyethylene glycol for this purpose; typical experiments involved the transfer of recombinant plasmids carrying a selectable marker gene, such as neomycin phosphotransferase (which confers resistance to kanamycin) linked to a suitable gene promoter, such as the nopaline synthase promoter. Genetically transformed protoplasts were then selected by their ability to survive and then divide to form a callus or tumour-like growth on media containing kanamycin. Transformation frequencies were very low, often as low as 0.001%. However, in more recent experiments, cells have been made leaky by electroporation (a high-voltage pulse of electric current) or by electroporation plus polyethylene glycol. Using these techniques, transformation frequencies of around 10% have been achieved with tobacco protoplasts (e.g. Shillito, Paszowski & Potrykus, 1985). Further, for plants such as tobacco, *Petunia* and sunflower, it has proved possible to grow kanamycin-resistant plants from the kanamycin-resistant cells. The NPT gene is integrated into the plant's nuclear DNA and is inherited by progeny. This then is a clear instance of successful genetic engineering. Further, the system is readily modified to incorporate the targeting and developmental regulatory mechanisms described earlier.

One of the exciting features of the direct DNA delivery system is that it does not rely on an infection. The limited host range of other vector delivery systems is therefore irrelevant, and the way is opened for genetic engineering of cereals. Cereal protoplasts are equally amenable to uptake of foreign DNA after electroporation and the system therefore has potential for use with the major crop species. However, there is at present one drawback, namely that for cereals it has not yet proved possible to grow fertile whole plants from the genetically transformed cells.

It will have become obvious to the reader that genetic engineering of monocotyledonous plants, including cereals, is presenting major problems, and that novel approaches are required to solve these problems. One recent approach has been the direct injection of foreign DNA (as before, a plasmid carrying the NPT gene linked to the *nos* promoter) directly into male developing flowering shoots (de la Pena, Lorz & Schell, 1987). This clearly results in the genetic transformation of some of the pollen cells, since after crossing with other injected plants, some of the progeny are kanamycin resistant. The proportion is low (under 0.1%) but, nevertheless, the system is obviously worthy of further investigation.

Finally, the use of shoot meristems as recipient tissues for *Agrobacterium*-mediated transformation (Ulian *et al.*, 1988) seems to us (and indeed to Ulian *et al.*) to have potential for monocots. Electroporated, genetically manipulated monocot protoplasts have not yet been grown into fertile transgenic plants and, faced with such problems, practitioners of clonal propagation would turn to meristems as source material. If therefore, a system can be developed for transfer of foreign DNA to monocot meristems, either using a 'forced' infection by *Agrobacterium* or using direct uptake of DNA in some sort of vectorless system which does not require preparation of protoplasts, then a major step towards routine genetic manipulation of cereals will have been taken.

### Application of the tools of molecular biology/genetics
*Reverse genetics*

Having isolated and characterised a gene which we think may be of adaptive significance during stress, we now have the possibility of reintroducing the gene to a stress-susceptible host plant to see whether it confers stress tolerance. If it does, we may safely conclude that the gene is of adaptive significance. This powerful approach has been called 'reverse genetics', having been originally devised as a tool for the analysis of gene structure and function (see Schell, 1987). It is applied as follows: we start with a purified gene, and produce defined mutations *in vitro*; we then reintroduce the gene to the host and then ask questions about the new phenotype, rather than working 'forwards' from the phenotype as in conventional genetics (Fig. 4). From the point of view of stress tolerance, the approach may be used both as originally intended, i.e. to analyse the structure–function relationships of genes known to be of adaptive significance (both structural and regulatory genes), as will be discussed later, but also to demonstrate adaptive significance, and further as a tool for gene identification. In the latter case, for instance, we might be in the position of knowing that a particular physiological aspect of tolerance segregates or correlates with the presence of a particular chromosome or chromosome arm, which similarly segregates or correlates with actual stress tolerance. By random cloning of sequences from the isolated chromosome coupled to reverse genetics, one could precisely identify the genetic region responsible for stress tolerance and confirm that it also confers the physiological trait originally observed. Thus one would demonstrate adaptive significance

Fig. 4. A comparison of the approaches of reverse genetics and conventional genetics for the analysis of gene function.

REVERSE GENETICS - - - - *V.* - - - - GENETICS

| GENETIC FUNCTION | heritable trait | GENETIC FUNCTION |
|---|---|---|
| ↓ | | ↓ |
| ISOLATE GENE | | NATURAL VARIATION |
| ↓ | | ↓ |
| MANIPULATE | | INDUCED MUTATION |
| ↓ | | ↓ |
| TRANSPLANT | | INTERBREED |
| ↓ | | ↓ |
| CONSEQUENCES? | | CONSEQUENCES? |

adaptive?

while paving the way for identification of the gene responsible, by its nucleotide sequence and subsequently in terms of the gene product it specifies. However, because of the number of random clones involved in working with a chromosome or chromosome arm, it is likely that the approach will only be of value where the stress tolerance can be assessed on cultured cells since the regeneration and screening of large numbers of plants from transformed cells represents too enormous a task.

As this suggests, the major problem we face is the recognition and isolation of genes which confer stress tolerance. Thereafter, we may use reverse genetics to confirm the adaptive significance of these genes. Other approaches to gene isolation will now be discussed.

### The cDNA route to gene isolation

Gene isolation is rendered most simple if there is available a corresponding gene product (protein) for which we have some sequence information or an antibody. The approach is further facilitated if the product is abundant and is therefore represented by an abundant mRNA

Fig. 5. Methods used for isolation and analysis of genes for which a product (protein or mRNA) can be identified.

GENE ISOLATION AND ANALYSIS

| | |
|---|---|
| GENE FUNCTION | |
| ↓ | |
| PROTEIN | amino acid sequence data/antibody |
| ↓ | |
| mRNA | from specialised organ or stressed plant |
| ↓ | |
| cDNA LIBRARY | SELECTION |
| ↓ | |
| cDNA CLONE | sequence analysis/comparisons |
| ↓ | |
| GENOMIC CLONE | sequence analysis/comparisons intron/exon organisation, flanking sequences |
| ↓ | |
| REGULATORY ELEMENTS | promoters, enhancers |
| ↓ | |
| RELATED GENES | sequence homology, common promoters |
| ↓ | |
| REVERSE GENETICS | novel host, promoter/reporter-gene constructs |

species within the mRNA population (Fig. 5). The first step, the construction of a cDNA library of cloned sequences representing the entire mRNA population, is now a routine step for which commercial kits are available. The cDNA clone of the gene of interest is isolated by selective hybridisation of a synthetic oligonucleotide probe (representing a region of amino acid sequence of the gene product). Alternatively a cDNA expression vector system may be used and the clone of interest identified using the corresponding antibody. Successful isolation of a cDNA clone is generally followed by sequence analysis to confirm its identity (i.e. that it encodes a protein of the correct size). It may then be used in turn as a probe to isolate the corresponding genomic clone from a library of clones constructed using $\lambda$ phage or cosmid vectors. Sequence comparisons between genomic clones and cDNA clones can be used to gain information about the gene in terms of, for example, intron/exon structure, and transcriptional activation signals. Examination of flanking sequences over a longer range will lead to the identification of regulatory elements (promoters and enhancers), as described earlier in the chapter. Comparison of the derived amino acid sequences of the gene product with the collection of sequences in one of the data banks may lead to a functional identification of the gene.

### Gene isolation by linkage

In some sequences we may have a genetic understanding of stress tolerance to the point where we know that it, or some aspect of it, appears to segregate as a single gene. In this case, although there is little reason for

Fig. 6. The potential of, and problems with, gene isolation by linkage.

GENE ISOLATION BY LINKAGE

GENETIC FUNCTION        heritable trait
        ↓

⌈  DEMONSTRATION OF LINKAGE TO AN ACCESSIBLE GENE

or ⟨

⌊  DEMONSTRATION OF LINKAGE TO AN 'R F L P'
        (restriction fragment length polymorphism)

        ↓

CHROMOSOME WALKING

        DIFFICULTY

HOW TO RECOGNISE THE GENE?

                REVERSE GENETICS?

HOW TO ANALYSE ITS FUNCTION?

using reverse genetics to demonstrate adaptive significance, it is possible, in principle, to isolate the gene in question by linkage to a polymorphic locus which has already been cloned (Fig. 6). Such a locus may either be defined by an allelomorph or by a restriction fragment length polymorphism (RFLP) in order for genetic linkage to be established. The extensive collections of RFLPs (Tanksley, 1983) being made for many crop species will facilitate the latter approach. The technique for isolating the gene in question depends upon 'walking' along the genome, via a set of overlapping clones (isolated from a genomic library by successive rounds of cross-hybridisation), from the cloned polymorphic gene to the target gene. This process has intrinsic difficulties for plants because of dispersed reiterated sequences which can present false overlaps and, of course, when we are dealing with an unknown gene it may be difficult to know when we have reached it. Reverse genetics may offer a solution to the latter problem.

Having considered the nature of the tools, we must now look at some routes for their application.

### Relevance to stress tolerance

The increasing ability to manipulate plants genetically is relevant to stress tolerance in two ways: first, there is the ability to analyse gene regulatory sequences relevant to stress tolerance and secondly, there is the hope that suitable genes may be transferred to economically important species in order to improve their stress tolerance. In relation to the first point, it has already been mentioned that particular genes may be put into plants in such a way that the genes are active in specific cells, tissues or organs. This is achieved by making a recombinant gene construct containing specific enhancer sequences. As has been pointed out by Schell (1987), the ability to manipulate plants genetically is a powerful tool in detection and analysis of sequences of this type. In respect of stress tolerance, for example, it is highly probable that genes which are transcribed only under conditions of stress will be under the control of a stress-regulated enhancer (see Ho & Sachs, Chapter 9). The splicing of a putative stress-regulated enhancer to a reporter gene such as kanamycin resistance or $\beta$-glucuronidase followed by delivery of the recombinant DNA to a suitable host then provides a means of testing the putative stress-regulated sequence. All that needs to be done is to expose the transgenic plant to the appropriate stress and assay for the activity of the reporter gene. The isolation of such stress-regulated enhancer sequences is an essential step in manipulating plants for stressful environments. The other major step is, of course, to identify the genes which confer tolerance and these are the subjects of the remainder of this chapter.

The potential power of recombinant DNA and associated techniques as research tools for molecular genetics has been extrapolated by many to the genetic manipulation of crop plants. In consequence there has been a general assumption that this form of genetic manipulation will contribute to or even supersede conventional plant breeding, with occasional extreme claims for a new revolution in agriculture. In order to retain some objectivity it is important to remember that the techniques described in the earlier parts of this chapter are at present limited to the transfer of single or small numbers of genes and also by our current level of understanding of the relationship between identifiable genes and plant performance, which is rather restricted. This is particularly true for plant stress and stress tolerance, which is an important goal for plant improvement. However, it is timely to consider the contribution which molecular genetics can make to the analysis and manipulation of genes conferring stress tolerance, given the constraints imposed by the small number of genes which may be simultaneously manipulated.

The potential of the approach is clearly seen in the production of transgenic herbicide-resistant and insect-resistant plants (della-Cioppa *et al.*, 1987; Vaeck *et al.*, 1987). The success of both these examples depends on an understanding of the mechanism of resistance and also on the fact that single genes are involved. As is emphasised by Blum (Chapter 11) and by Yeo & Flowers (Chapter 12), resistance to the major stresses reducing crop production, i.e. drought, temperature extremes and salinity, depends on many genes possibly with interacting effects. Against this background we will concentrate on drought and examine how the above approaches may be applied.

Drought is perhaps one of the most complex examples to choose but it illustrates well the possibilities of, and pitfalls to, progress. Drought affects almost every facet of plant function and we are faced with the paradox that yield and evapotranspiration are intimately linked. In general, increases in yield when water supply is limiting are likely to result from characteristics which increase the available water supply, increase water use efficiency or increase biomass allocation to the economically useful plant parts (Passioura, 1986). Additionally, features which maintain cell viability and protect metabolism in water-stressed tissue and allow rapid recovery after dry periods will contribute yield under some circumstances.

### Analysis of the stress response
#### Stress-induced proteins
A significant observation has been that under stress conditions new families of proteins are synthesised. This holds for temperature extremes,

Table 1. *Examples of increases in mRNA and specific proteins brought about by drought (D), salinity (S) or polyethylene glycol (PEG)*

| System | Product | Reference |
|---|---|---|
| **mRNA increases** | | |
| Mesocotyl of maize seedlings (D) | mRNA hybridising with HSP70 genomic clone | Heikkila *et al.*, 1984 |
| Barley leaves (D) | α-amylase mRNA and protein | Jacobsen, Hanson & Chandler, 1986 |
| *Mesembryanthemum crystallinum* leaves (S) | PEP carboxylase mRNA | Ostrem *et al.*, 1987 |
| Barley leaves and roots (S) | Unidentified mRNAs | Ramagopal, 1987 |
| Wheat leaves and roots (S) | Unidentified mRNAs | Gulick & Dvorak, 1987 |
| **Protein increases** | | |
| Barley leaves (D) | 60 kDa protein | Dasgupta & Bewley, 1984 |
| Tobacco cell cultures (S and PEG) | 26 kDa protein 'Osmotin' | Ericson & Alfinito, 1984; Singh *et al.*, 1987a |
| *Mesembryanthemum crystallinum* leaves (S) | PEP carboxylase | Höfner *et al.*, 1987 |
| *Brassica napus* roots (D) | Unidentified proteins | Vartanian, Damerval & de Vienne, 1987 |
| Soybean stem cell walls (D) | 28 kDa protein | Bozarth, Mullet & Boyer, 1987 |

drought, salinity and anoxia, and is discussed in detail by Ho & Sachs (Chapter 9).

During drought, overall rates of protein synthesis decline (Dhindsa & Cleland, 1975; Bewley & Larsen, 1980). Nevertheless, some specific proteins and mRNA species increase in amount in droughted plants. Some examples are given in Table 1. The functions of many of these proteins have not been established. The types formed seem to differ according to species, tissue type and growth conditions and so far no universal set of proteins equivalent to the heat-shock proteins (Chapter 9) has been found. It must be said, however, that the genes directing synthesis of the major stress-induced proteins have proved particularly amenable to analysis via the cDNA cloning route. We wait with interest to see whether the stress-response proteins are adaptive in the sense that the capacity to synthesise them at an augmented level or the synthesis of unique members of the family differentiate between stress-sensitive and stress-tolerant plants.

Reverse genetics will be powerful in the analysis of the control of the stress response starting with a structural and functional comparison of the promoter and enhancer elements which regulate the synthesis of stress-response proteins. This will lead, as has already been demonstrated for the heat-shock proteins (Chapter 9), to the identification of common sequence elements involved in the transformation of stress signals into gene activity, and to an eventual understanding of how stresses such as water deficit are perceived at the cellular level. An important element in this analysis is the opportunity of fusing promoter and enhancer sequences (Fig. 7) to readily detectable reporter genes such as the $\beta$-glucuronidase gene (GUS) (Jefferson, Kavanagh & Bevan, 1987). The fusion of a soybean heat-shock promoter to the GUS gene has already been achieved (R.A. Jefferson, personal communication) and the chimeric construction functions as expected during heat shock when reintroduced into plants. The availability of transgenic plants containing such constructs will provide a powerful tool for the analysis of cellular hierachies in the stress response.

A further opportunity for the use of stress-responsive promoters and enhancers is as probes to isolate other stress-responsive genes, the activity of which is not manifest by protein synthesis. As regards the manipulation of stress tolerance as a breeding tool, it is likely that the stress-responsive promoters and enhancers will have a role to play in controlling the expression of adaptive genes when these are transplanted over great evolutionary distances.

Fig. 7. The use of stress-inducible promoter and enhancer sequences.

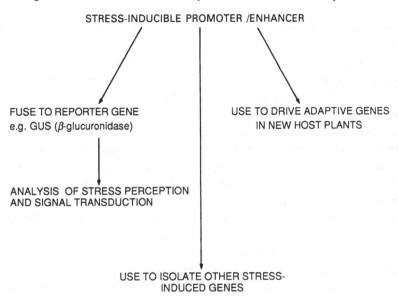

STRESS-INDUCIBLE PROMOTER /ENHANCER

FUSE TO REPORTER GENE
e.g. GUS ($\beta$-glucuronidase)

USE TO DRIVE ADAPTIVE GENES
IN NEW HOST PLANTS

ANALYSIS OF STRESS PERCEPTION
AND SIGNAL TRANSDUCTION

USE TO ISOLATE OTHER STRESS-
INDUCED GENES

Table 2. *Molecules and enzymes with possible protective roles during*
*drought*

| | |
|---|---|
| Ascorbate | Betaine |
| Glutathione | Proline |
| α-tocopherol | Polyols (mannitol, sorbitol, pinitol) |
| Superoxide dismutase | Dehydroascorbate reductase |
| Glutathione reductase | Monodehydroascorbate reductase |
| Ascorbate peroxidase | Catalase |

*Protective metabolism during stress*

Ability to survive and recover from stresses such as drought (and temperature extremes) may involve maintenance of structural and metabolic integrity. Some heat-shock proteins could have this function (Chapter 9). It is possible that during drought plants may be susceptible to oxidative and free radical damage (Price & Hendry, 1987; Smirnoff & Colombé, 1988). A number of systems are known which can protect against this type of damage and these can be induced during drought (Price & Hendry, 1987; Smirnoff & Colombé, 1988). Enzymes involved in hydrogen peroxide scavenging, glutathione reductase, ascorbate peroxidase and monodehydroascorbate reductase (Table 2) increase in activity during drought (Smirnoff & Colombé, 1988).

Accumulation of compatible solutes (glycine betaine, proline and polyols such as mannitol, sorbitol and pinitol) occurs in many droughted plants and they act as cytoplasmic osmotica for osmotic adjustment. However, they may have other functions which include enhancing the stability of macromolecules and membranes (Paleg, Stewart & Starr, 1985; Smirnoff & Stewart, 1985; Chapter 7).

Although all the functions suggested above are not firmly proven, they are highlighted because they illustrate an approach different from that in the previous section, i.e. to identify and understand important physiological processes during drought and then to seek to enhance them. The processes mentioned above, compared with growth and development, are relatively simple and could be amenable to manipulation. The enzymes involved in betaine synthesis have not been characterised yet but are under active study (Hanson & Grumet, 1985; Weigel, Lerma & Hanson, 1988).

Considering the example of drought-induced increase in monodehydroascorbate reductase activity (Fig. 8) and assuming the increased activity reflects increased synthesis, isolation of the corresponding genes via the cDNA route should give us access to another set of stress-inducible promoter elements. Furthermore, it is possible that further potentiating this type of response by reintroduction of the corresponding genes coupled to more powerful stress-responsive promoters or enhancers could generate greater stress tolerance in the recipient plants.

### Abscisic acid

It is almost obligatory to mention the growth regulator abscisic acid (ABA) which accumulates during drought and appears to influence many plant responses including stomatal closure (Chapter 5) and osmotic adjustment (Chapters 5 and 6; La Rosa *et al.*, 1987). Although ABA generally inhibits protein synthesis, it can increase the synthesis of specific proteins during salinity (Singh *et al.*, 1987b) and drought (Heikkila *et al.*, 1984). Since ABA accumulation itself requires transcription (Guerrero & Mullet, 1986) the question arises as to what actually is the signal (and how is it transduced) which leads to this change in gene transcription.

Isogenic lines of spring wheat differing in capacity for ABA accumulation

Fig. 8. The relationship between leaf water potential ($\psi_L$) and mono-dehydroascorbate reductase activity in barley leaves during a 7-day drying period. $r = 0.72$, $P<0.001$ (from Smirnoff & Colombé, 1988).

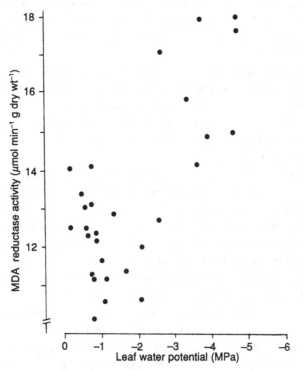

have been produced by conventional breeding techniques and the high ABA lines have better yield under dry conditions (Innes, Blackwell & Quarrie, 1984). ABA-deficient and insensitive mutants of a number of species are also available (Walton, 1980).

Clearly the genes, yet to be identified, which control ABA synthesis will be of interest and should offer another class of stress-inducible promoter and enhancer elements. Some intense biochemistry and protein chemistry lie ahead for those who undertake this gene cloning exercise via the cDNA route.

### The use of isogenic plants for genes conferring stress tolerance

This method, based on conventional breeding techniques, allows the production of lines differing in expression of an inherited character in similar genetic backgrounds (Austin, 1987; Richards, 1987). The advantage of this character can then be tested, with the caveat that its effect may depend on the genetic background employed. Table 3 lists some of the characters which have been tested in this way. Since genetic variability for these characters exists, a further use for molecular genetics is questionable. This is particularly the case for complex characters such as xylem vessel diameter because their expression is not influenced by drought. The remaining characters are of greater interest in this context because their expression is enhanced by drought (see references in Table 3). These isogenic lines may provide an opportunity to study controls over the expression of genes and this would be relevant for abscisic acid and glycinebetaine where controls over synthesis and turnover are not fully understood, but where rate-controlling enzymes might occur (Hanson & Grumet, 1985).

Another approach is to select tolerant and sensitive lines from a reasonably genetically uniform population. In either case there is an opportunity to compare protein profiles from the tolerant and sensitive lines by two-dimensional gel electrophoresis. Protein differences may then represent the genetic differences and lead one, via protein chemistry and the cDNA cloning route, to the gene(s) conferring tolerance. The route is fraught with difficulties, especially at the level of recognition of genuine protein differences which at best may be only quantitative. Alternatively the extensive genetic analysis which has been exploited in the construction of isogenic lines may be extended to the demonstration of linkage to RFLPs followed by 'walking' to the gene. The practicality of such an approach will be limited to those instances where we are dealing with a single locus for a particular aspect of stress tolerance which can be recognised in isolation. Thus, if we combine this with the inherent difficulties of genomic walking, it

Table 3. *Examples of characters with possible roles in drought resistance for which isogenic lines or isopopulations have been developed*

| Character | Mechanism | Effectiveness | References |
|---|---|---|---|
| Narrow xylem vessels in wheat seminal roots | Decreased hydraulic conductance | + | Richards, 1987 |
| Glaucousness (epicuticular wax) | Water-use efficiency, leaf temperature and duration | + | Johnson, Richards & Turner, 1983 |
| Osmotic adjustment (low solute potential) | Cell expansion and soil water extraction at lower water potential | +<br>– * | Morgan & Condon, 1986<br>Grumet, Albrechtson & Hanson, 1987 |
| Abscisic acid accumulation during drought | Water-use efficiency? | + | Innes *et al.*, 1984 |

*The low solute potential ($\psi_s$) isopopulation was selected by high glycine betaine concentration.

is expected that tractable systems for molecular genetics will be the exception.

### The use of plants from extreme environments

Wild plants from extreme environments may possess genes and gene combinations which confer stress tolerance. We must realise, however, that many of their characteristics, e.g. leaf pubescence and succulence in drought-resistant plants, are incompatible with the high yield potential required for crop plants. In addition, most of these species contain compounds such as phenolics and mucilages which interfere with conventional molecular biology techniques.

Some wild species have larger capacities for osmotic adjustment, a trait which may improve yield during drought (Table 3, Turner, 1986). Interesting examples of this are *Dubautia* species from Hawaii which differ in osmotic adjustment mainly as a result of differences in cell wall elasticity. Interspecific hybrids can be made which have intermediate properties (Robichaux, Holsinger & Morse, 1986). Material such as this could make a basis for the molecular study of differences in cell wall elasticity.

The example we will consider in more detail is that of inducible CAM (Crassulacean acid metabolism). CAM is a modification of photosynthesis found in succulent plants in which carbon dioxide is fixed at night by phospho*enol*pyruvate carboxylase (PEP carboxylase) and the final product, malic acid, is stored in vacuoles. During the day it is decarboxylated and the carbon dioxide refixed by $C_3$ photosynthesis. It is considered to be a drought adaptation (Ting, 1985).

There is a small group of succulent plants which carry out $C_3$ photosynthesis when well watered but switch to CAM when subjected to drought or salinity. These include *Mesembryanthemum crystallinum* (Holtum & Winter, 1982) and *Sedum telephium* (Groenhof, Bryant & Etherington, 1986; Lee & Griffiths, 1987). In *M. crystallinum* the $C_3$–CAM switch involves large increases in the extractable activities of nine enzymes involved in CAM including PEP carboxylase, NAD and NADP–malic enzymes, and pyruvate phosphate dikinase (Holtum & Winter, 1982). A tonoplast ATPase, possibly associated with malic acid transport into the vacuole, also increases in activity when CAM is induced (Struve *et al.*, 1985). It is possible that all these enzymes are coordinately regulated and switched on by drought or salinity. In *M. crystallinum* the increased PEP carboxylase activity is a result of an increase in the amount of enzyme protein by *de novo* synthesis (Höfner *et al.*, 1987) which in turn results from an increase in translatable mRNA (Ostrem *et al.*, 1987). In *Sedum telephium* induction of CAM, shown by malate fluctuations, occurs within 24 h of imposing water stress and increases in PEP carboxylase activity also occur in droughted plants (Groenhof, Bryant & Etherington, 1988; A.C. Groenhof, N. Smirnoff and J.A. Bryant, unpublished).

CAM is unlikely to be useful for any of the conventional food crops. Nevertheless, further study of those plants in which CAM is inducible could prove useful since isolation of, for example, the PEP carboxylase gene will permit the isolation of its controlling sequences. This will provide us with another set of stress-specific (drought) promoters and enhancers. Furthermore the PEP carboxylase gene has proved amenable to cloning via the cDNA route (Harpster & Taylor, 1986).

### Conclusions

Techniques are now well established for transferring foreign genes into certain dicotyledonous species, and there is the prospect of extending these techniques to useful crop species. These techniques provide powerful means both of analysing plant genes and of manipulating plants in a heritable manner. Applying this technology to improving the stress-tolerance of crop species will involve a study of the whole range of biochemical and physiological responses to stress, in order to identify those responses which are amenable to the molecular genetic approach, i.e. which are mediated by single genes or by a small group of genes. It will also be necessary to determine which responses to stress are truly adaptive. The molecular genetic approach provides a means of doing this, and also provides a means of obtaining suitable gene-regulatory elements to ensure appropriately controlled transcription of stress-adaptive genes transferred into new host plants. Such studies are as yet in their very early stages, and it is necessary to achieve a much greater understanding of basic plant biochemistry and physiology in order for molecular biology to be used usefully in improving plant stress tolerance.

### References
Austin, R.B. (1987). Some crop characteristics of wheat and their influences on yield and water use. In *Drought Tolerance in Winter Cereals*, ed. J.P. Srivastava, E. Porceddu, E. Acevedo and S. Varma, pp. 321–36. Chichester: John Wiley.

Bewley, J.D. & Larsen, K. (1980). Cessation of protein synthesis in water-stressed pea roots and maize mesocotyls without loss of polyribosomes. Effects of lethal and non-lethal water stress. *Journal of Experimental Botany*, **32**, 1245–56.

Bozarth, C.S., Mullet, J.E. & Boyer, J.S. (1987). Cell wall proteins at low water potentials. *Plant Physiology*, **85**, 261–7.

Bryant, J.A. (1988). Putting genes into plants. *Plants Today*, **1**, 23–8.

Dasgupta, J. & Bewley, J.D. (1984). Variations in protein synthesis in different regions of greening leaves of barley seedlings and effects of imposed water stress. *Journal of Experimental Botany*, **35**, 1450–9.

de la Pena, A., Lorz, H. & Schell, J. (1987). Transgenic rye plants obtained by injection of DNA into young floral tillers. *Nature*, **325**, 274–6.

della-Cioppa, G., Bauer, S.C., Taylor, M.L., Rochester, D.E., Klein, B.K., Shah, D.M., Fraley, R.T. & Kishore, G.M. (1987). Targeting a herbicide-resistant enzyme from *Escherichia coli* to chloroplasts of higher plants. *Bio/technology*, **5**, 579–84.

Dhindsa, R.S. & Cleland, R.E. (1975). Water stress and protein synthesis. I. Differential inhibition of protein synthesis. *Plant Physiology*, **55**, 778–81.

Ericson, M. & Alfinito, S.H. (1984). Proteins produced during salt stress in tobacco cell culture. *Plant Physiology*, **74**, 506–9.

Grimsley, N., Hohn, T., Davies, J.W. & Hohn, B. (1987). *Agrobacterium*-mediated delivery of infectious maize streak virus into maize plants. *Nature*, **325**, 177–9.

Groenhof, A.C., Bryant, J.A. & Etherington, J.R. (1986). Photosynthetic changes in the inducible CAM plant *Sedum telephium* L. following the imposition of water stress. I. General characteristics. *Annals of Botany*, **57**, 689–95.

Groenhof, A.C., Bryant, J.A. & Etherington, J.R. (1988). Photosynthetic changes in the inducible CAM plant *Sedum telephium* L. following the imposition of water stress. II. Changes in the activity of phosphoenolpyruvate carboxylase. *Annals of Botany*, **62**, 187–92.

Grumet, R., Albrechtson, R.S. & Hanson, A.D. (1987). Growth and yield of barley isopopulations differing in solute potential. *Crop Science*, **27**, 991–5.

Guerrero, F. & Mullet, J.E. (1986). Increased abscisic acid biosynthesis during plant dehydration requires transcription. *Plant Physiology*, **80**, 588–91.

Gulick, P. & Dvorak, J. (1987). Gene induction and repression by salt treatment in roots of the salinity-sensitive Chinese Spring Wheat and salinity-tolerant Chinese Spring × *Elytrigia elongata* amphiploid. *Proceedings of the National Academy of Sciences, USA*, **84**, 99–103.

Hanson, A.D. & Grumet, R. (1985). Betaine accumulation: metabolic pathways and genetics. In *Cellular and Molecular Biology of Plant Stress*, ed. J.L. Key and T. Tosuge, pp. 71–92. New York: Alan R. Liss.

Harpster, M.H. & Taylor, W.C. (1986). Maize phosphoenolpyruvate carboxylases. Cloning and characterization of mRNAs encoding isozymic forms. *Journal of Biological Chemistry*, **261**, 6132–6.

Heikkila, J.J., Papp, J.E.T., Schultz, G.A. & Bewley, J.D. (1984). Induction of heat shock protein messenger RNA in maize mesocotyls by water stress, abscisic acid and wounding. *Plant Physiology*, **76**, 270–4.

Hernalsteens, J.O., Thia-Toong, L., Schell, J. & Van Mongtagu, M. (1985). An *Agrobacterium*-transformed cell culture from the monocot *Asparagus officinalis*. *EMBO Journal*, **3**, 3039–41.

Herrera-Estrella, L., Van den Broek, G., Maenhault, R., Van Montagu, M., Schell, J., Timko, M. & Cashmore, A. (1985). Light-inducible and chloroplast-associated expression of a chimaeric gene introduced into *Nicotiana tabacum* using a Ti plasmic vector. *Nature*, **310**, 115–20.

Höfner, R., Vazquez-Morena, L., Winter, K., Bohnert, H.J. & Schmitt, J.M. (1987). Induction of Crassulacean acid metabolism in *Mesembryanthemum crystallinum* by high salinity: mass increase and *de novo* synthesis of PEP-carboxylase. *Plant Physiology*, **83**, 915–19.

Holtum, J.A.M. & Winter, K. (1982). Activity of enzymes of carbon metabolism during the induction of Crassulacean acid metabolism in *Mesembryanthemum crystallinum* L. *Planta*, **155**, 8–16.

Hooykaas-van Slogteren, G.M.S., Hooykaas, P.J.J. & Schilperoort, R.A. (1984). Expression of Ti plasmid genes in monocotyledonous plants infected with *Agrobacterium tumefaciens*. *Nature*, **311**, 763–4.

Innes, P., Blackwell, R.D. & Quarrie, S.A. (1984). Some effects of genetic variation in drought induced abscisic acid accumulation on the yield and water use of spring wheat. *Journal of Agricultural Science, Cambridge*, **102**, 341–51.

Jacobsen, J.V., Hanson, A.D. & Chandler, P.C. (1986). Water stress enhances expression of an α-amylase gene in barley leaves. *Plant Physiology*, **80**, 350–9.

Jefferson, R.A., Kavanagh, T.A. & Bevan, M.W. (1987). Gus fusions: β-glucuronidase as a sensitive and versatile gene marker in higher plants. *EMBO Journal*, **6**, 3901–7.

Johnson, D.A., Richards, R.A. & Turner, N.C. (1983). Yield, water relations and surface reflectances of near-isogenic wheat lines differing in glaucousness. *Crop Science*, **23**, 318–25.

La Rosa, P.C., Hasegawa, P.M., Rhodes, D., Clithero, J.M., Watas, A.A. & Bressan, R.A. (1987). Abscisic acid stimulated osmotic adjustment and its involvement in adaptation of tobacco cells to NaCl. *Plant Physiology*, **85**, 174–81.

154    *S.G. Hughes* et al.

Lee, H.S.J. & Griffiths, H. (1987). Induction and repression of CAM in *Sedum telephium* L. in response to photoperiod and water stress. *Journal of Experimental Botany*, **38**, 834–41.
Morgan, J.M. & Condon, A.G. (1986). Water use, grain yield and osmoregulation in wheat. *Australian Journal of Plant Physiology*, **13**, 523–32.
Ostrem, J.A., Olson, S.W., Schmitt, J.M. & Bohnert, H.J. (1987). Salt stress increases the level of translatable mRNA for phosphoenolpyruvate carboxylase in *Mesembryanthemum crystallinum*. *Plant Physiology*, **84**, 1270–5.
Paleg, L.G., Stewart, G.R. & Starr, R. (1985). The effects of compatible solutes on protein. *Plants and Soil*, **89**, 83–94.
Passioura, J.B. (1986). Resistance to drought and salinity: avenues for improvement. *Australian Journal of Plant Physiology*, **13**, 191–201.
Price, A. & Hendry, G.A.F. (1987). The significance of the tocopherols in stress survival in plants. In *Free Radicals, Oxidant Stress and Drug Action*, ed. C. Rice-Evans, pp. 443–5. London: Richelieu Press.
Ramagopal, S. (1987). Differential mRNA transcription during salinity stress in barley. *Proceedings of the National Academy of Sciences, USA*, **84**, 94–8.
Richards, R.A. (1987). Physiology and the breeding of winter-grown cereals for dry areas. In *Drought Tolerance in Winter Cereals*, ed. J.P. Srivastava, E. Porceddu, E. Acevedo and S. Varma, pp. 133–50. Chichester: John Wiley.
Robichaux, R.H., Holsinger, K.E. & Morse, S.R. (1986). Turgor maintenance in Hawaiian *Dubautia* species: the role of variation in tissue osmotic and elastic properties. In *The Economy of Plant Form and Function*, ed. T. Givnish, pp. 353–80. Cambridge: Cambridge University Press.
Schell, J. (1987). Transgenic plants on tools to study the molecular organization of plant genes. *Science*, **237**, 1176–83.
Shillito, R.D., Paszowski, J. & Potrykus, I. (1985). High efficiency gene transfer to plants. *Bio/Technology*, **3**, 1099–103.
Singh, N.K., Bracker, C.A.S., Hasegawa, P.M., Handa, A.K., Bruckel, S., Hermondson, M.A., Pfankoch, E., Regnier, F.E. & Bressan, R.A. (1987*a*). Characterisation of osmotin. A Thaumatin-like protein associated with osmotic adaptation in plant cells. *Plant Physiology*, **85**, 528–36.
Singh, N.K., La Rosa, P.C., Handa, A.K., Hasegawa, P.M. & Bressan, R.A. (1987*b*). Hormonal regulation of protein synthesis associated with salt tolerance in plant cells. *Proceedings of the National Academy of Sciences, USA*, **84**, 739–43.
Smirnoff, J. & Colombé, S.V. (1988). Drought influences the activity of enzymes of the chloroplast hydrogen peroxide scavenging system. *Journal of Experimental Botany*, **39**, 1097–1109.
Smirnoff, N. & Stewart, G.R. (1985). Stress metabolites and their role in coastal plants. *Vegetatio*, **62**, 273–8.
Struve, I., Weber, A., Lüttge, U., Ball, E. & Smith, J.A.C. (1985). Increased vacuolar ATPase activity correlated with CAM induction in *Mesembryanthemum crystallinum* and *Kalanchoe blossfeldiana* cv. Tom Thumb. *Journal of Plant Physiology*, **117**, 451–68.
Tanksley, S.D. (1983). Molecular markers in plant breeding. *Plant Molecular Biology Reporter*, **1**, 3–8.
Ting, I.P. (1985). Crassulacean Acid Metabolism. *Annual Review of Plant Physiology*, **36**, 595–622.

Turner, N.C. (1986). Adaptation to water deficits: a changing perspective. *Australian Journal of Plant Physiology*, **13**, 175–90.

Ulian, E.C., Smith, R.H., Gould, J.H. & McKnight, T.D. (1988). Transformation of plants via the shoot apex. *In Vitro Cell and Developmental Biology*, **24**, 951–4.

Vaeck, M., Reynaerts, A., Höfte, H., Jansens, S., De Beukeleer, M., Dean, C., Zabeau, M., Van Montagu, M. & Leemans, J. (1987). Transgenic plants protected from insect attack. *Nature*, **328**, 33–7.

Van den Broek, G., Timko, M.P., Kausch, A.P., Cashmore, A.R., Van Montagu, M. & Herrera-Estrella, L. (1985). Targeting of a foreign protein to chloroplasts by fusion to the transit peptide from the small subunit of ribulose 1,5-biphosphate carboxylase. *Nature*, **313**, 358–63.

Vartanian, N., Damerval, G. & de Vienne, D. (1987). Drought-induced changes in protein patterns of *Brassica napus* var. *oleifera* roots. *Plant Physiology*, **84**, 989–92.

Walden, R. (1988). *Genetic Transformation in Plants*. Milton Keynes, UK: Open University Press.

Walton, D.C. (1980). The biochemistry and physiology of abscisic acid. *Annual Review of Plant Physiology*, **31**, 453–89.

Watson, J.D., Tooze, J. & Kurtz, D.T. (1983). *Recombinant DNA: a short course*. New York: W.H. Freeman.

Weigel, P., Lerma, C. & Hanson, A.D. (1988). Choline oxidation by intact spinach chloroplasts. *Plant Physiology*, **86**, 54–60.

TUAN-HUA DAVID HO AND
MARTIN M. SACHS

# 9 Environmental control of gene expression and stress proteins in plants

## Introduction

Plants are constantly subject to adverse environmental conditions such as drought, flooding, extreme temperatures, excessive salts, heavy metals, high-intensity irradiation and infection by pathogenic agents. Because of their immobility, plants have to make necessary metabolic and structural adjustments to cope with the stress conditions. To this end, the expression of the genetic programme in plants is altered by the stress stimuli to induce and/or suppress the production of specific proteins which are either structural proteins or enzymes for specific metabolic pathways.

Several problems are addressed in the study of stress-induced proteins: (1) Perception: how does a plant recognise the existence of a stressful condition? (2) Regulation of gene expression: how does the perceived stress signal alter the expression of genes? (3) Function: what are the physiological roles of the stress-induced proteins? Studies designed to answer these questions usually begin with the finding of new proteins in stressed tissues, most likely by gel electrophoretic techniques. This initial observation is followed by purification of the stress proteins, and the cloning and characterisation of their genes. Research on the function of stress proteins has been progressing, although many stress proteins remain unidentified. The least understood process is probably the molecular mechanism underlying the perception of stress signals.

## Temperature stress

Due to seasonal changes, almost all plants are affected by temperature fluctuations in their life cycles. Very high temperatures have been reported in many arid zones around the world, and the lack of effective transpiration in plants located in these areas causes the temperatures inside these plants to be significantly higher than ambient (Levitt, 1980). Chilling or subfreezing temperatures are even more common. In temperate zones, most plants can encounter a wide temperature range, from higher than 40 °C in summer to subfreezing temperatures in winter. A rapid temperature upshift on a sunny winter morning, or a downshift in temperature as would occur after sunset in a desert, makes the condition even more stressful. Three types of adjustments are expected in temperature-related stress conditions. First, macromolecules such as proteins are denatured at high temperatures; thus, denatured proteins have to be removed and their

Fig. 1. SDS gel analysis of proteins synthesised by excised maize roots incubated at continuous 40 °C. Roots of 3-day-old maize seedlings were excised and incubated at 40 °C for increasing times as indicated. Labelling with [$^{35}$S]methionine was carried out in the final 20 min of the incubation. Proteins were visualised by fluorography. Mol wt distribution in kDa indicated at left. From Cooper & Ho (1983).

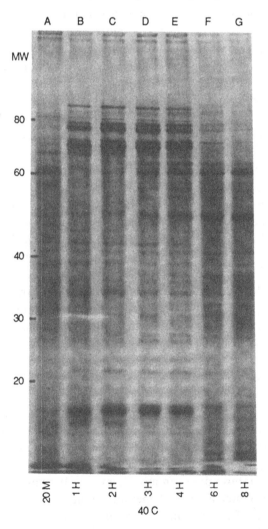

replacements synthesised. Second, metabolic pathways are often affected by temperature perturbations, causing the accumulation or depletion of certain metabolites. Third, some physical properties of membranes such as lipid fluidity are influenced by temperature shifts.

### Heat-stress induced gene expression

The most readily observable response to heat stress in many organisms is the induction of heat shock proteins (HSPs) (for reviews, see Schlesinger, Ashburner & Tissieres, 1982; Craig, 1986). This phenomenon was first investigated in the fruit fly *Drosophila melanogaster*. When *Drosophila* cells are rapidly shifted from their normal growth temperature (25 °C) to an increased temperature (37 °C), there is a cessation of normal protein synthesis with the concomitant synthesis of a novel set of HSPs. This alteration of gene expression is accompanied by the regression of old polytene chromosome puffs and the generation of new ones (for review see Ashburner & Bonner, 1979).

The induction of HSPs has been studied in several higher plants including soybean, pea, tobacco, tomato and maize (for review see Sachs & Ho, 1986). Besides the common size groups of HSPs (such as HSP70), higher plants also synthesise a unique group of small HSPs with a size of 15–18 kDa (Key, Lin & Chen, 1981; Cooper & Ho, 1983; Fig. 1). All the HSPs appear to be coordinately expressed when the tissue is under heat stress. The optimal condition for HSP induction in higher plants is a drastic temperature upshift to 39–42 °C. However, a drastic temperature surge (shock) is not absolutely required for HSP induction; these proteins are also induced if there is a gradual temperature rise such as a 2.5 °C h$^{-1}$ increase, a condition closer to that which occurs in the field (Altschuler & Mascarenhas, 1982). Furthermore, HSPs and their mRNAs have been detected in field-grown plants under heat stress (Burke *et al.*, 1985; Kimpel & Key, 1985). Thus it is more appropriate to call these proteins 'heat stress proteins' rather than the commonly used 'heat shock proteins', although the abbreviation 'HSP' remains the same. HSP synthesis can be detected within 20 min of heat stress, and the increase in transcript levels of some *Hsp* genes is noted within 3–5 min (Schöffl & Key, 1982; Fig. 2). However, the induction of HSP appears to be transient, lasting for only a few hours despite the continuous presence of heat-stress temperatures (Schöffl & Key, 1982; Cooper & Ho, 1983; Fig. 2).

The induction of HSPs correlates with the increase of the levels of their transcripts. Liquid hybridisation studies conducted by Schöffl & Key (1982) using cloned cDNA probes have revealed that about 20 different species of HSP18 mRNA accumulate to 19 000 copies per cell within two hours of heat stress in soybean cells. When a heat-stressed tissue is returned to normal temperatures (e.g. 28 °C) the synthesis of HSPs decreases over the next few hours with a concomitant decline in the levels of gene transcripts. Several plant *Hsp* genes have been cloned. The maize *Hsp70* gene appears to be very similar to the *Drosophila* counterpart, with 75% sequence homology in the coding region (Rochester, Winter & Shah, 1986). The three soybean *Hsp18* genes share greater than 90% homology in their deduced amino acid sequences. They also share similarities with the *Drosophila* small HSPs (22–27 kDa) in hydropathy profiles (Schöffl, Raschke & Nagao, 1984). Besides the TATA box, sequences related to the *Drosophila* heat-shock consensus regulatory element (CTnGAAnnTTCnAG; Pelham & Bienz, 1982) are found − 48 to − 62 base pairs 5′ to the start of transcription (Schöffl *et al.*, 1984). In addition, there are secondary heat-shock consensus elements located further upstream (Nagao *et al.*, 1985). The DNA sequence analysis of these plant *Hsp* genes supports the view that the molecular mechanism involved in the induction of *Hsp* genes is highly conserved among euka-ryotes. To analyse further the promoter regions, Gurley *et al.* (1986) have introduced a soybean small *Hsp* gene into primary sunflower tumors via Ti plasmid-mediated transformation. They found that a region from − 95 to − 192 in the 5′ flanking region of the soybean *Hsp* gene is important for the thermoinducible transcription in the transformed sunflower cells.

Fig. 2. Time course of accumulation of HSP mRNA. One μg of poly(A)⁺RNA isolated from soybean hypocotyls after different times of incubation at 42.5 °C (hs) or at additional times after transfer back to 28 °C after 4 h at the elevated temperature (recovery), were electrophoresed in formaldehyde agarose gels. Blots of these gels were hybridised with a mixture of four cDNAs encoding small soybean HSPs ranging from 15 to 23 kDa. From Schöffl & Key (1982).

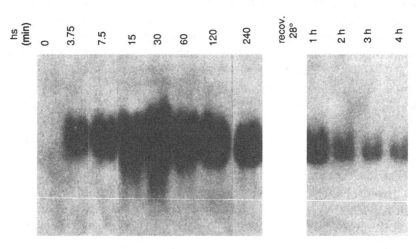

Although it is not clear how heat stress causes the induction of HSPs, a transcription factor which binds to the promoter of *Hsp* has been studied in *Drosophila* cells (Weiderrecht *et al.*, 1987; Wu *et al.*, 1987). Besides heat stress, many other factors, including osmotic stress, high salt, 2,4-dinitrophenol, arsenite, anaerobiosis and high concentrations of abscisic acid (ABA), ethylene or auxins, also induce the synthesis of certain plant HSPs (Czarnecka *et al.*, 1984). In soybean, arsenite and cadmium induce a normal spectrum of HSPs, yet some of the factors induce HSP27 specifically. Although the physiological significance of these inductions is not known, these factors can certainly be used as additional tools to investigate the molecular mechanisms underlying *Hsp* gene expression.

The induction of HSPs is accompanied by an alteration in the expression of other genes. In *Drosophila*, HSPs are essentially the only proteins that are synthesised during heat stress. The expression of normal proteins is suppressed and many pre-existing mRNAs are sequestered rather than being degraded (Storti *et al.*, 1980). When returned to a normal temperature the synthesis of normal proteins resumes even in the absence of new transcription (Lindquist, 1981). Thus, while the induction of HSPs in *Drosophila* is likely to be at the transcriptional level, the repression of the synthesis of normal proteins is at the translational level. In yeast, the pre-existing mRNAs are not sequestered but undergo normal turnover (Lindquist, 1981). Hence, in the absence of continuous synthesis of these pre-existing mRNAs their levels decrease gradually during heat stress. In higher plants, the effect of heat stress on the synthesis of normal proteins appears to be diverse. Vierling & Key (1985) have examined the effect of heat stress on the synthesis of ribulose 1,5 bisphosphate carboxylase in a soybean cell culture line. The synthesis of both the large and small subunits of this enzyme is decreased by 80% during heat stress. The levels of mRNA for the small subunit (encoded in the nuclear genome) also decrease during heat stress. In contrast, changes in the synthesis of the large subunit (encoded in the chloroplast genome) show little relationship to the corresponding mRNA levels; large subunit mRNA levels remain relatively unchanged by heat stress. Belanger, Brodl & Ho (1986) have found that in barley aleurone layers, the synthesis of secretory proteins such as $\alpha$-amylase and protease is preferentially suppressed by heat stress. The mRNAs encoding these secretory proteins are normally very stable, yet they are quickly degraded during heat stress. The degradation of these mRNAs correlates with heat-stress-induced destruction of endoplasmic reticulum (ER). Since mRNAs encoding secretory proteins are normally translated by ER-bound polyribosomes, it is suggested that the heat-stress-induced ER destruction leads to the selective degradation of mRNA for secretory proteins. In contrast, a mRNA encoding a non-secretory protein, actin, remains stable during heat stress (Ho *et al.*, 1987). In developing *Phaseolus* seeds, Chrispeels & Greenwood (1987) have shown that although the synthesis of phytohaemagglutinin is enhanced, the post-translational processing of this protein is retarded during heat stress. It is not yet known why the synthesis of certain proteins is preserved while the synthesis of other proteins is quickly suppressed by heat stress.

### Potential function of HSPs

Although the function of most of the HSPs in plants has not been elucidated, some of these proteins have been identified in other organisms: ubiquitin in chicken embryo fibroblasts (Bond & Schlesinger, 1985), lysyl tRNA synthetase (Hirschfield *et al.*, 1981) and ATP-dependent protease in *E. coli* (Philips, Van Bogelen & Neidhardt, 1984), an isozyme of enolase in yeast (Iida & Yahara, 1985), and the uncoating ATPase that releases triskelia from coated vesicles in mammals (Ungewickell, 1985). Other properties of HSPs include the binding of these proteins to poly(A)$^+$ RNA in Hela cells (Schönfelder, Horsch & Schmid, 1985), to fatty acid in rat (Guidon & Hightower, 1986), and to collagen in chicken embryo fibroblasts (Nagata, Saga & Yamada, 1986). It has been shown that the induction of HSPs allows the cells to establish thermotolerance in many organisms (Schlesinger *et al.*, 1982), i.e. to survive at a temperature that is normally lethal. Lin, Roberts & Key (1984) have shown that briefly subjecting soybean seedlings to 40 °C followed by incubation at 28 °C results in the induction of HSPs and a concomitant establishment of thermotolerance. All the tissues in maize plants, with the exception of germinating pollen, synthesise HSPs during heat stress (Cooper, Ho & Hauptmann, 1984; Xiao & Mascarenhas, 1985). It is known that germinating pollen is more sensitive to high temperature than other tissues (Herrero & Johnson, 1980), a phenomenon that also suggests that the ability to synthesise HSPs is related to thermotolerance. However, this view is questioned by the observation that it is possible to establish thermotolerance in germinating *Tradescantia* pollen without the induction of HSPs (Xiao & Mascarenhas, 1985).

Some plant HSPs are known to be associated with chloroplasts. Although chloroplasts and mitochondria do not synthesise HSP themselves (Nieto-Sotelo & Ho, 1987), certain nuclear encoded HSPs synthesised in the cytosol have been shown to be transported into chloroplasts (Kloppstech *et al.*, 1985; Vierling *et al.*, 1986). The HSP22 of *Chlamydomonas* is incorporated into the thylakoid membrane without size reduction, while in pea, HSP22 is synthesised as a 26 kDa precursor. In *Chlamydomonas*, Schuster *et al.* (1988) have shown that the HSP22 is associated with the photosystem

II complex. The algal cells become more resistant to photoinhibition at high light intensity if they have been subjected to prior heat-shock treatment. Blocking the synthesis of HSPs by the cytosolic protein synthesis inhibitor, cycloheximide, eliminates the protection against photoinhibition. Thus, the association of HSPs with photosystem II is likely to be related to the establishment of photoinhibition. It has been shown that HSPs are also associated with mitochondria, ribosomes and nuclei, yet the significance of these associations is not yet clear (Lin *et al.*, 1984; Cooper & Ho, 1987). In tomato cells, it has been observed that HSPs form 30–50 nm wide aggregates termed the heat-shock granules (Nover, Scharf & Neumann, 1983). Since specific sets of mRNA are found in the heat-shock granules, it is conceivable that these granules are formed to sequester normal mRNA whose translation is blocked during heat stress (L. Nover, K.D. Scharf & D. Neumann, personal communication). One of the best known functions of HSPs is probably the involvement of ubiquitin in proteolysis. Ubiquitin is a 76 amino-acid-long protein which is induced by heat stress in both animals and plants (Bond & Schlesinger, 1985; M.R. Brodl, M.L. Tierney & T.-H.D. Ho, unpublished). The carboxylic terminus of ubiquitin can link with the $\epsilon$-amino group of lysine residues in other proteins, and most of the ubiquitinated proteins are then degraded by a special protease system. It is likely that during heat stress, ubiquitin will tag the thermally denatured proteins for degradation; thus new proteins can be formed after the stress is over.

### Cold temperature-induced gene expression

Low, non-freezing, temperatures are essential for important physiological processes such as vernalisation, stratification, and cold acclimation for freezing tolerance. Low temperatures have been shown to cause changes in protein content, enzyme activities, and membrane structures (Levitt, 1980). The increase of freezing tolerance of plants during cold acclimation has been suggested to be the consequence of physiological and metabolic changes dependent on gene expression (Weiser, 1970). Although cold-induced proteins have been observed by gel electrophoretic analysis (for example see Guy, Niemi & Brambl, 1985), the identity and function of these proteins remain unclear.

### Drought and salt stress-induced proteins

Several physiological changes induced by drought have been documented, including an increase in ABA levels, the closure of stomata and the increase in cellular osmolarity. The increase in ABA levels is probably due to the *de novo* synthesis of this hormone, and the process requires transcription of certain genes (Guerrero & Mullet, 1986). The increase in ABA appears to be related to the closure of stomata to avoid further water loss. To date, the enzyme system involved in ABA biosynthesis has not been well investigated. The most commonly observed cellular components enhanced by drought are proline and betaine. A 10- to 100-fold increase of free proline content occurs in leaf tissues of many plants during moderate water deficit (Hanson & Hitz, 1982). It is not clear, however, whether changes in enzyme activity or gene expression are involved in the increase in proline. The increase in betaine is probably more related to salinity stress. One of the enzymes involved in the biosynthesis of betaine has been purified (A.D. Hanson, personal communication), which should facilitate the investigation into the molecular mechanisms underlying the regulation of betaine synthesis. In bacterial cells, it has been shown that DNA supercoiling plays a crucial role in the transcription of a gene encoding an osmotically inducible betaine transport system (Higgins *et al.*, 1988). It is not yet known whether a similar mechanism operates in higher plants. The elevated levels of proline and betaine during drought could probably serve as cytosolic osmotica.

It has also been reported that osmotic shock enhances the biosynthesis of polyamines (Flores & Galston, 1982). However, the concentrations of polyamines seem to be too low to account for any significant increase in osmolarity.

Drought also has a profound effect on protein synthesis. In many plant tissues, a reduced water potential causes a reduction of total protein synthesis and a rapid dissociation of polyribosomes. The latter has been shown not to be the consequence of increase in ribonuclease activity (Hsiao, 1973; Dhindsa & Bewley, 1976). For a specific protein, Jacobsen, Hanson & Chandler (1986) have shown in barley leaves that water stress enhances the synthesis of one of the $\alpha$-amylase isozymes. Using a cDNA probe they found that water-stressed leaves contained much more $\alpha$-amylase mRNA than unstressed plants.

Low water potential has a profound effect on chloroplast activities, including the decreases in electron transport and photophosphorylation (Mayoral *et al.*, 1981) and the changes in conformation of the thylakoid and of the coupling factor (Younis, Boyer & Govindjee, 1979). Some of these effects are similar to exposing thylakoid or coupling factor to $Mg^{2+}$ above 5 mM (Younis, Weber & Boyer, 1983), thus leading to the suggestion that an increase in $Mg^{2+}$ concentration in chloroplasts due to water loss could cause these changes (Kaiser, Schröppel-Meier & Wirth, 1986).

Salinity is another major limiting factor in agriculture affecting a large area of cultivated land, and with increasing irrigation salinity stress has become more widespread. In recent years there has been increased interest in the introduction of halophytes and in the breeding of salt-resistant crop plants to be used in coastal areas and lands that require heavy irrigation. Many culture cell lines adapted to high salt conditions have been isolated. To study the biochemical mechanism of salt adaptation, Singh *et al.* (1985) have examined the protein profiles in a NaCl-adapted (salt-adapted) line of *Nicotiana tabacum* and a non-NaCl-adapted line. They found an abundant 26 kDa protein among several new or enhanced protein bands with increasing levels of NaCl adaptation, while the intensities of a few other protein bands were reduced. This 26 kDa protein has a pI greater than 8.2, and it is also induced by ABA whose level is enhanced in salt-adapted cells. However, the synthesis of ABA-induced 26 kDa protein is transient unless the cultured cells are simultaneously exposed to NaCl stress. Singh *et al.* (1987) suggest that ABA is involved in the normal induction of the synthesis of this 26 kDa protein and that the presence of NaCl is necessary for the protein to accumulate. In tobacco plants many tissues synthesise the 26 kDA protein in response to ABA treatment, but the highest level of expression of this protein was observed in the outer stem tissue. Although exogenously applied ABA induces the synthesis of an immunologically cross-reactive 26 kDa protein in cultured cells of several plant species, the physiological role of this 26 kDa protein remains unknown. King, Hussey & Turner (1986) found that cells of tobacco cultured in suspension accumulate this protein even in the absence of NaCl as the cells approach the stationary phase, but the accumulation never reaches the level seen in the salt-adapted cells. This protein can also be induced by other agents that lower the water potential, such as PEG and KCl, but no increase in the levels of this protein is seen after heat stress or heavy metal ($CdCl_2$) treatment. Although this protein accumulates with salt stress in hydroponically grown tomato plants, the levels of this protein do not seem to correlate with natural salt tolerance in wild tomato species. Therefore, it is speculated that the 26 kDa protein plays a role in responding to lowered water potential in plants instead of being related to salt tolerance. The induction of new mRNAs has also been reported in salt-stressed monocots. For example, in wheat, salt stress induces ten new mRNAs and suppresses eight mRNAs in root tissue, but not in shoot tissues (Gulick & Dvorak, 1987).

An intriguing stress-induced alteration in gene expression occurs in a succulent plant, *Mesembryanthemum crystallinum*, which switches its primary photosynthetic $CO_2$ fixation pathway from $C_3$ type to CAM (Crassulacean acid metabolism) type upon salt or drought stress (Winter, 1974; Chapter 8). Ostrem *et al.* (1987) have shown that the pathway switching involves an increase in the level of mRNA encoding phosphoenol-pyruvate carboxylase, a key enzyme in CAM photosynthesis.

### Anaerobic stress

Anaerobic treatment results in a drastic alteration in the pattern of protein synthesis in maize seedlings (Sachs, Freeling & Okimoto, 1980). Pre-existing (aerobic) protein synthesis is repressed while selective synthesis of new polypeptides is initiated (Sachs et al., 1980). This is most likely a plant's natural response to flooding.

Studies on the maize anaerobic response stemmed from the extensive analysis of the maize ADH (alcohol dehydrogenase) system by D. Schwartz and coworkers (reviewed in Freeling & Bennett, 1985; Gerlach et al., 1986). Initially it was shown that ADH activity in maize seedlings increases as a result of flooding (Hageman & Flesher, 1960). Freeling (1973) later reported that ADH activity increased at a zero order rate between 5 and 72 h of anaerobic treatment, reflecting a simultaneous expression of two unlinked genes, Adh1 and Adh2. Schwartz (1969) showed that ADH activity is required to allow the survival of maize seeds and seedlings during flooding. ADH is the major terminal enzyme of fermentation in plants and is responsible for recycling $NAD^+$ during anoxia. It has been suggested that ethanolic fermentation permits tight cytoplasmic pH regulation, thus preventing acidosis from competing lactic fermentation (Roberts et al., 1984; Roberts, Andrade & Anderson, 1985).

Except for one possible overlap, anaerobiosis induces a different set of proteins in maize than heat stress (Kelley & Freeling, 1982). In maize seedlings, as is the case for many heat stress systems (cf. Schlesinger et al., 1982), repression of pre-existing (aerobic) protein synthesis and the induction of a new set of proteins occurs very shortly after being subjected to anoxia (e.g. an argon atmosphere: Sachs et al., 1980). As mentioned above, the soybean HSP28 is induced by a number of different environmental insults, including both heat stress and anaerobiosis (Czarnecka et al., 1984).

The induction of anaerobic treatment synthesis occurs in two phases. During the first few hours of anaerobic treatment there is a transition period during which there is a rapid increase in the synthesis of a class of polypeptides with an approximate molecular weight of 33 kDa (Fig. 3). These have been referred to as the transition polypeptides (TPs) as they represent most of the protein synthesis occurring in early anaerobiosis.

After approximately 90 min of anoxia, the induced synthesis of an additional group of c. 20 polypeptides is first detected. This group of 20 anaerobic polypeptides (ANPs) represents more than 70% of the total labelled amino acid incorporation after five hours of anaerobiosis (Fig. 3). By this time the synthesis of the TPs is at a minimal level; however, these polypeptides accumulate to a high level during early anaerobiosis and have been shown by pulse-chase experiments to be very stable. The synthesis of the ANPs continues at a constant rate for up to *c.* 72 h of anaerobic treatment (depending on which maize line is being examined), at which time protein synthesis decreases concurrently with the onset of cell death (Sachs *et al.*, 1980). It has been shown in maize primary roots that hypoxia causes the induction of the ANPs, but does not cause the complete repression of pre-existing protein synthesis (Kelley & Freeling, 1982). In addition, a novel set of polypeptides, not normally observed under aerobic or anaerobic conditions, is synthesised under hypoxic conditions (Kelley & Freeling, 1982).

Fig. 3. Protein synthesis in a maize primary root during: (*a*) one hr pulse labelling with [³H]leucine under aerobic conditions; (*b*)–(*e*) pulse labelling with [³H]leucine during the specified times under anaerobic conditions. The arrow labelled 'TPs' indicates the position of the transition polypeptides. The unlabelled arrow indicates the position of alcohol dehydrogenase 1 (ADH1). From Sachs *et al.* (1980).

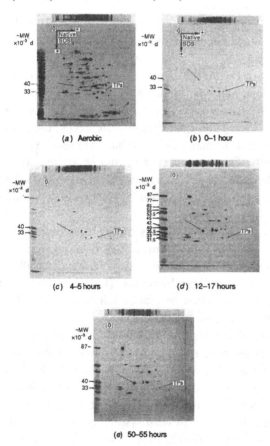

The identities of some of the ANPs are now known. The isozymes of alcohol dehydrogenase, encoded by the *Adh1* and *Adh2* genes, have been identified as ANPs through the use of genetic variants (Sachs & Freeling, 1978). More recently, glucose 6 phosphate isomerase (Kelley & Freeling, 1984a), fructose 1,6 diphosphate aldolase (Kelley & Freeling, 1984b), and sucrose synthase (Hake *et al.*, 1985) have been identified as ANPs. Pyruvate decarboxylase activity has also been shown to be induced by anaerobiosis (Laszlo & St Lawrence, 1983) and therefore is probably one of the ANPs. The identities and functions of the remaining ANPs or of the TPs are unknown. It may be noted, however, that five of the ANPs so far identified are glycolytic enzymes, and sucrose synthase is also involved in glucose metabolism. In the light of the inability of maize seedlings to survive five hours of flooding in the absence of an active *Adh1* gene (Schwartz, 1969), it appears that at least one function of the anaerobic response is to enable the plant to produce as much ATP as possible when there is an oxygen deficit, as would occur in waterlogged soils.

In the presence of air, the roots, coleoptile, mesocotyl, endosperm, scutellum, and anther wall of maize synthesise a tissue-specific spectrum of polypeptides. The scutellum and endosperm of the immature kernel synthesise many or all of the ANPs constitutively, along with many other proteins under aerobic conditions. Under anaerobic conditions all of the above organs selectively synthesise only the ANPs. Moreover, except for a few characteristic qualitative and quantitative differences, the patterns of anaerobic protein synthesis in these diverse organs are remarkably similar (Okimoto *et al.*, 1980). On the other hand, maize leaves, which have emerged from the coleoptile, do not incorporate labelled amino acids under anaerobic conditions and do not survive even a brief exposure to anaerobiosis (Okimoto *et al.*, 1980).

The shift in pattern of protein synthesis during anaerobiosis has been observed in root tissue of many other plant species including rice, sorghum, barley, pea, and carrot (see Sachs & Ho, 1986). In anaerobically treated barley aleurone cells, lactate dehydrogenase (LDH) activity increases (Hanson & Jacobsen, 1984) as does enzyme activity and mRNA levels for ADH (Hanson, Jacobsen & Zwar, 1984).

The rapid repression of pre-existing protein synthesis caused by anaerobic treatment is correlated with a near complete dissociation of polysomes in primary roots of soybeans (Lin & Key, 1967) and maize (E.S. Dennis and A.J. Pryor, personal communication). This does not result from degradation of 'aerobic' mRNAs, because the mRNAs encoding the pre-existing proteins remain translatable in an *in vitro* system at least five hours after anaerobic treatment is initiated (Sachs *et al.*, 1980). This is in agreement with the observation that the polysomes, dissociated by anaerobiosis, rapidly reform up to 80–90% of their pretreatment levels, even in the absence of new RNA synthesis, when soybean seedlings are returned to air (Lin & Key, 1967).

The molecular basis of the maize anaerobic response has been analysed with cDNA clones from high molecular weight poly(A)$^+$ RNA of anaerobically treated maize seedling roots (Gerlach *et al.*, 1982). Cloned anaerobic-specific cDNAs were identified by colony hybridisation analysis, using differential hybridisation to labelled cDNA of mRNA from anaerobic and aerobic roots. The anaerobic-specific cDNA clones were grouped into families on the basis of cross-hybridisation, and several of the families were analysed by hybrid-selected translation and by RNA gel blot (Northern) hybridisation. The *Adh1* and *Adh2* cloned cDNA families were subsequently identified from this anaerobic-specific cDNA clone library, and the cDNA clone families and the genes encoding them were analysed extensively (Dennis *et al.*, 1984, 1985; Sachs *et al.*, 1986).

The anaerobic-specific cDNA clones were used as probes to measure gene expression in maize seedling roots and shoots. In both tissues, the levels of mRNA hybridisable to the cDNAs increase during anaerobic treatment. This has been quantified in the case of *Adh1* and *Adh2*, with the kinetics of mRNA increase being the same for both mRNAs. The mRNA level first begins to increase after 90 min of anaerobic treatment, and continues to increase until it reaches a 50-fold above the aerobic level after five hours of anaerobiosis. This level is maintained until after 48 h in the case of *Adh1* but starts declining after 10 h in the case of *Adh2* (Dennis *et al.*, 1985). This pattern of mRNA level increase and decrease is reflected in the previously described rates of *in vivo* anaerobic protein synthesis for ADH1 and ADH2 (Sachs *et al.*, 1980). Sucrose synthase (ANP87) also appears to have the same kinetics of mRNA increase (Hake *et al.*, 1985). *In vitro* run-off transcription experiments (Rowland & Strommer, 1985; L. Beach, personal communication) show that in root cells, there is an increase in the transcription rate of the *Adh1* gene during anoxia, indicating that the increase in the levels of anaerobic-specific mRNAs is due to induced transcription of the *Anp* genes.

A comparison of the regions of the *Adh1* and *Adh2* genes upstream from the site of transcription initiation reveals only a few islands of homology. These include an 11 bp region that includes the 'TATA box' and three additional 8 bp regions of homology. One or more of these sequences might account for the anaerobic control of these genes (Dennis *et al.*, 1985). *In vitro* mutagenesis coupled with transformation and gene expression studies, as well as the analysis of the 5' regions of the other anaerobic genes, will be necessary to determine which if any of these homologous regions might be important in regulating the induced expression of the anaerobic response genes.

Using electroporation as a transformation technique together with *in vitro* mutagenesis of the *Adh1* promoter, Walker *et al.* (1987), found two regions upstream from the site of transcription initiation which appear to control the anaerobic induction of mRNA accumulation. Sequences homologous to those detected by Walker *et al.* (1987) are found in the *Adh1* and *Adh2* genes of maize and an *Adh1*-like gene from pea (Llewellyn *et al.*, 1987). However, there are no obviously comparable regions upstream of the *Adh* gene of *Arabidopsis thaliana* (Chang & Meyerowitz, 1986). Transformation by electroporation of DNA into maize protoplasts is a promising general system for *in vitro* analysis of gene expression; but this system has drawbacks. First, maize protoplasts cannot yet be regenerated into intact plants. Secondly, the cell culture systems used to make protoplasts for these studies are derived from maize embryo (scutellum), a tissue where ADH1 and other ANPs are synthesised at a high constitutive level under aerobic conditions.

Ellis *et al.* (1987) reported a similar study using the *Agrobacterium* system to introduce the maize *Adh1* promoter, associated with a marker gene, into tobacco. In this case, some incompatibilities were found in expression of the maize (a monocot) promoter in tobacco (a dicot), as the maize promoter was functional in transgenic tobacco plants only when it was augmented with an enhancer-like sequence from a gene constitutively expressed in tobacco. It would be preferable to use a homologous maize system where intact transgenic plants could be obtained. Also it is necessary to test the expression of anaerobically stimulated promoters in tissues where they are normally inducible by anoxic treatment (e.g. the seedling root and shoot).

### Response to ultraviolet light exposure

Plants can be damaged by the ultraviolet rays of intense sunlight. As a defence, plants produce flavonoid pigments that absorb the UV irradiation, minimising damage to them. Cell suspension cultures of parsley produce and accumulate flavonoids when irradiated with ultraviolet (Hahlbrock, 1981). It was found that UV treatment causes the coordinate induction of phenylalanine ammonia-lyase, 4-coumarate:CoA ligase, chalcone synthase, and UDP-apiose synthase, all enzymes required for flavonoid biosynthesis. It was shown, using cDNA probes, that the levels of mRNAs encoding these enzymes increase during UV treatment (Kuhn *et al.*, 1984) and that this is the result of increased transcription rates of the genes (Chappel & Hahlbrock, 1984).

### Heavy metal-induced proteins and peptides

Heavy metals such as cadmium and copper contaminate the environment because of increasing activities in mining and industrial waste disposal. Most of these heavy metals are toxic to plant metabolism, and plants have developed both a strategy of avoiding uptake of these toxic metal ions and an ability to synthesise proteins and peptides that can tightly bind and sequester these metals. In *Datura innoxia*, resistance to cadmium is correlated with the ability to synthesise one or more cysteine-rich, metal binding peptides (Jackson *et al.*, 1985). Similar Cd-binding peptides have been isolated from several other plant species (Rauser, 1984; Reese & Wagner, 1987). Although some similarities exist, these plant Cd-binding peptides appear to be different from metallothioneins, the mammalian heavy metal-binding proteins (Webb, 1979). Grill, Winnacker & Zenk (1985) have elucidated the structure of a group of cysteine-rich peptides capable of binding heavy metal ions via thiolate coordination. These peptides, named phytochelatins, have a general structure of $[\gamma\text{Glu-Cys}]_n$-Gly ($n = 2$–8). Phytochelatins are induced by a wide range of metal ions including $Cd^{2+}$, $Zn^{2+}$, $Pb^{2+}$, $Ag^+$, $AsO_4^{3-}$ and $SeO_3^{2-}$. In cultured cells of several plant species, the synthesis of phytochelatins is induced within an hour after the addition of 200 $\mu M$ $Cd(NO_3)_2$. Kinetic studies indicate that phytochelatins are probably synthesised from glutathione or its precursor, $\gamma$-glutamylcysteine in a sequential manner, generating the set of homologous peptides (Grill, Winnacker & Zenk, 1987). It is apparent that small peptides such as phytochelatins are more efficient than metallothioneins in sequestering heavy metal ions.

### Biological stress

Besides the physical stress conditions discussed above, plants are constantly subjected to invasion by insects and animals and infection by microorganisms. To cope with these biological stresses, plants produce specific proteins or metabolites that discourage the invading agents from further damaging the rest of the plants. One of the best examples of this type of response is the accumulation of serine proteinase inhibitor proteins in the leaves of plants from the Solanaceae and Leguminosae familes when severely damaged by attacking insects or other mechanical agents (Ryan *et al.*, 1985). Cell wall fragments released from the damaged tissue can trigger a systemic response in other parts of plants (Ryan *et al.*, 1985). In response to this 'wound hormone' the plant expresses two small gene families of serine proteinase inhibitors. The accumulation of these inhibitors constitutes a defensive response that interferes with the digestive process of attacking agents.

Bishop *et al.* (1984) have isolated a pectic polysaccharide (6 kDa) from tomato leaves that can elicit the induction of proteinase inhibitors. Upon further hydrolysis this pectic polysaccharide yields oligogalacturanan (degree of polymerisation from 2 to 6) that still possesses the proteinase inhibitor inducing activities.

Mechanical wounding also induces the synthesis of HRGP (hydroxy-proline-rich glycoproteins) which are deposited in cell walls (Showalter *et al.*, 1985), such proteins apparently making the walls more rigid. The recent discovery that the seed coat is a rich source of HRGP suggests a protective role for these proteins against mechanical damage to the seeds (Cassab *et al.*, 1985).

In addition to forming passive physical barriers such as cuticles and rigid cell walls, there are at least two other means by which plants defend themselves against fungal infection: the infected plants produce specific secondary metabolites (phytoalexins) which are toxic to the invading fungus (for review see Bailey & Mansfield, 1982), or the plants synthesise enzymes which can hydrolyse fungal cell walls to stop the further migration of fungal hyphae. Most of the phytoalexins are derivatives of phenylalanine (Hahl-brock *et al.*, 1985), and phenylalanine ammonia-lyase and two other

Fig. 4. Comparison of the two signal-reaction chains leading either to the UV light-induced formation of flavonoids or to the elicitor-induced for-mation of furanocoumarins and related compounds with antimicrobial activity. From Hahlbrock *et al.* (1985). 'PR'-proteins are pathogenesis-related proteins.

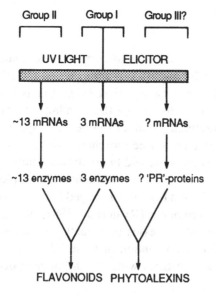

enzymes in the same pathway, chalcone synthase and chalcone isomerase, can be induced not only by infection but also by the fungal cell wall elicitors (Lawton *et al.*, 1983). Similar to UV light, the fungal elicitors induce the synthesis of three enzymes (phenylalanine ammonia-lyase, cinnamate 4-hydroxylase and 4-coumarate:CoA ligase) to convert phenylalanine to 4-coumaroyl-CoA, a key intermediate of the phenylpropanoid metabolism in higher plants. Thereafter, the induction patterns diverge in accordance with the metabolic end product (Hahlbrock *et al.*, 1985; Fig. 4). Using thiouridine incorporation to isolate newly synthesised RNA from elicitor-treated cells, Cramer *et al.* (1985) have demonstrated that the induction of these enzymes is part of rapid and extensive changes in the pattern of mRNA synthesis directing production of a set of proteins associated with expression of disease resistance.

Fungal elicitors can also induce cell wall degrading enzymes such as chitinase and $\beta$-1,3 glucanase, which damages the fungal cells. Chappell, Hahlbrock & Boller (1984) have shown that fungal elicitors can induce ethylene and the ethylene forming enzyme, 1-aminocyclopropane-1-carboxylic acid (ACC) synthase, several hours before the induction of chitinase. It is also demonstrated that exogenously applied ethylene can induce both chitinase and $\beta$-1,3 glucanase. However, the action of elicitors in the induction of these enzymes is unlikely to be mediated by ethylene because the potent ethylene biosynthesis inhibitor, aminoethoxyvinylglycine (AVG), has no effect on the elicitor-induced chitinase (Boller, 1985). It seems that ethylene and fungal infection appear to be separate, independent stimuli for the induction of these enzymes.

In addition to the proteins mentioned above, there have been many reports of the induction of 'pathogenesis-related' proteins (PRPs) in a number of plant species infected by viruses, viroids, bacteria or fungi (Van Loon, 1985). It has been shown that in cultured parsley cells, increased transcription of two PRP genes occurs within a few minutes of exposure to fungal elicitors (Somssich *et al.*, 1986). Although several PRPs from tobacco have been purified and their cDNAs isolated (Hooft van Huijsduijnen, Van Loon & Bol, 1986), the exact action of these proteins remains unclear.

### Summary

Most of the stress proteins discussed in this account have a role either in helping the plants survive, or in minimising the effectiveness of the stress agent. In helping plants survive under stress conditions, the stress proteins perform the following functions: (1) maintenance of the basic metabolism in the stressed cell, e.g. the induction of ADH and some glycolysis enzymes in anaerobic stressed cells; (2) protection of cellular components from being damaged by the stressful condition, e.g. the association of HSPs with enzymes or organelles resulting in tolerance to thermodenaturation; (3) removal of damaged cellular components, e.g. the tagging of denatured proteins by ubiquitin for proteolysis in heat-stressed cells. Although the stressed cells are usually not metabolically active, the induction of stress proteins can keep the cells from being killed and the cells can recover once the stress condition is relieved. In minimising the effectiveness of the stress agents, the stress proteins take up a more active set of functions: (1) the physical blockage of entry of stress agents, such as the induction of cell wall HRGP and lignin synthesis by fungal elicitors; (2) the sequestration of stressful agents, as in the induction of phytochelatin to chelate heavy metal ions; (3) the impairment of the biological stress agents, such as the induction of proteinase inhibitors and enzymes capable of hydrolysing fungal cell walls, and the synthesis of phytoalexins.

For future research in this field, in addition to physiological and biochemical approaches, genetic analysis will be essential in the establishment of causal relationships between the induction of a stress protein and the establishment of tolerance to the stress condition. In most cases it is not difficult to detect the induction of new proteins during stress. However, the induction of new proteins does not necessarily establish stress tolerance; it may well be the consequence of damage caused by stress conditions. Thus, genetic mutants will be necessary to test the physiological role of a stress protein.

### Acknowledgements
The authors' research is supported by NSF: DCB 8702299 to T-H.D.H. and NIH: 5 R01 GM 34740 to M.M.S.

### References
Altschuler, M. & Mascarenhas, J.P. (1982). Heat shock proteins and effects of heat shock in plants. *Plant Molecular Biology*, 1, 103–15.
Ashburner, M. & Bonner, J.J. (1979). The induction of gene activity in *Drosophila* by heat shock. *Cell*, 17, 241–54.
Bailey, J.A. & Mansfield, J.W. (ed.) (1982). *Phytoalexins*. Glasgow: Blackie.
Belanger, F.C., Brodl, M.R. & Ho, T.-H.D. (1986). Heat shock causes destabilization of specific mRNAs and destruction of endoplasmic reticulum in barley aleurone cells, *Proceedings of the National Academy of Sciences, USA*, 83, 1354–8.
Bishop, P.D., Pearce, G., Bryant, J.E. & Ryan, C.A. (1984). Isolation and characterization of the protease inhibitor-inducing factor from tomato leaves, identity and activity of poly- and oligogalacturonide fragments. *Journal of Biological Chemistry*, 259, 13172–7.
Boller, T. (1985). Induction of hydrolases as a reaction against pathogens. In

*Cellular and Molecular Biology of Plant Stress*, ed. J.L. Key and T. Kosuge, pp. 247–62. New York: Academic Press.

Bond, U. & Schlesinger, M.J. (1985). Ubiquitin is a heat shock protein in chicken embryo fibroblasts. *Molecular and Cellular Biology*, **5**, 949–56.

Burke, J.J., Hatfield, J.L., Klein, R.R. & Mullet, J.E. (1985). Accumulation of heat shock proteins in field-grown cotton. *Plant Physiology*, **78**, 394–8.

Cassab, G.I., Nieto-Sotelo, J., Cooper, J.B., Van Holst, G.-J. & Varner, J.E. (1985). A developmentally regulated hydroxyproline-rich glycoprotein from the cell walls of soybean coats. *Plant Physiology*, **77**, 532–5.

Chang, C. & Meyerowitz, E.M. (1986). Molecular cloning and DNA sequence of the *Arabidopsis thaliana* alcohol dehydrogenase gene. *Proceedings of the National Academy of Sciences, USA*, **83**, 1408–12.

Chappel, J. & Hahlbrock, K. (1984). Transcription of plant defence genes in response to UV or fungal elicitor. *Nature*, **311**, 76–8.

Chappell, J., Hahlbrock, K. & Boller, T. (1984). Rapid induction of ethylene biosynthesis in cultured parsley cells by fungal elicitor and its relationship to the induction of phenylalanine ammonia-lyase. *Planta*, **161**, 475–80.

Chrispeels, M.J. & Greenwood, J.S. (1987). Heat stress enhances phytohemagglutinin synthesis but inhibits its transport out of the endoplasmic reticulum. *Plant Physiology*, **83**, 778–84.

Cooper, P. & Ho, T.-H.D. (1983). Heat shock proteins in maize. *Plant Physiology*, **71**, 215–22.

Cooper, P. & Ho, T.-H.D. (1987). Intracellular localization of heat shock proteins in maize. *Plant Physiology*, **84**, 1197–203.

Cooper, P., Ho, T.-H.D. & Hauptmann, R.M. (1984). Tissue specificity of the heat-shock response in maize. *Plant Physiology*, **75**, 431–41.

Craig, E.A. (1986). The heat shock responses. *CRC Review of Biochemistry*, **18**, 239–80.

Cramer, C.L., Ryder, T.B., Bell, J.N. & Lamb, C.J. (1985). Rapid switching of plant gene expression induced by fungal elicitor. *Science*, **227**, 1240–2.

Czarnecka, E., Edelman, L., Schöffl, F. & Key, J.L. (1984). Comparative analysis of physical stress responses in soybean seedlings using cloned heat shock cDNAs. *Plant Molecular Biology*, **3**, 45–58.

Dennis, E.S., Gerlach, W.L., Pryor, A.J., Bennetzen, J.L., Inglis, A., Llewellyn, D., Sachs, M.M., Ferl, R.J. & Peacock, W.J. (1984). Molecular analysis of the alcohol dehydrogenase (*Adh1*) gene of maize. *Nucleic Acids Research*, **12**, 3983–4000.

Dennis, E.S., Sachs, M.M., Gerlach, W.L., Finnegan, E.J. & Peacock, W.J. (1985). Molecular analysis of the alcohol dehydrogenase 2 (*Adh2*) gene of maize. *Nucleic Acids Research*, **13**, 727–43.

Dhindsa, R.S. & Bewley, J.D. (1976). Plant desiccation: polysome loss not due to ribonuclease. *Science*, **191**, 181–2.

Ellis, J.G., Llewellyn, D.J., Dennis, E.S. & Peacock, W.J. (1987). Maize *Adh-1* promoter sequences control anaerobic regulation: addition of upstream promoter elements from constitutive genes is necessary for expression in tobacco. *EMBO Journal*, **6**, 11–16.

Flores, H.E. & Galston, A.W. (1982). Polyamine and plant stress: activation of putrescine biosynthesis by osmotic shock. *Science*, **217**, 1259–60.

Freeling, M. (1973). Simultaneous induction by anaerobiosis or 2,4D or multiple enzymes specified by two unlinked genes: differential *Adh1–Adh2* expression in maize. *Molecular and General Genetics*, **127**, 215–27.

Freeling, M. & Bennett, D.C. (1985). Maize *Adh1*. *Annual Review of Genetics*, **19**, 297–323.

Gerlach, W.L., Pryor, A.J., Dennis, E.S., Ferl, R.J., Sachs, M.M. & Peacock, W.J. (1982). cDNA cloning and induction of the alcohol dehydrogenase gene (*Adh1*) of maize. *Proceedings of the National Academy of Sciences, USA*, **79**, 2981–5.

Gerlach, W.L., Sachs, M.M., Llewellyn, D., Finnegan, E.J. & Dennis, E.S. (1986). Maize *alcohol dehydrogenase*: A molecular perspective. In *Plant Gene Research Vol. 3, A genetic approach to biochemistry*, ed. A.D. Blonstein and P.J. King, pp. 73–100. Vienna: Springer-Verlag.

Grill, E., Winnacker, E.-L. & Zenk, M.H. (1985). Phytochelatins: The principal heavy-metal complexing peptides of higher plants. *Science*, **230**, 674–6.

Grill, E., Winnacker, E.-L. & Zenk, M.H. (1987). Phytochelatins, a class of heavy-metal-binding peptides from plants, are functionally analogous to metallothioneins, *Proceedings of the National Academy of Sciences, USA*, **84**, 439–43.

Guerrero, F. & Mullet, J.E. (1986). Increased abscisic acid biosynthesis during plant dehydration requires transcription. *Plant Physiology*, **80**, 588–91.

Guidon, P.T. & Hightower, L.E. (1986). Purification and initial characterization of the 71-kilodalton rat heat-shock protein and its cognate as fatty acid binding proteins. *Biochemistry*, **25**, 3231–9.

Gulick, P. & Dvorak, J. (1987). Gene induction and repression by salt treatment in roots of the salinity-sensitive Chinese Spring wheat and the salinity-tolerant Chinese Spring × *Elytrigia elonata* amphiploid. *Proceedings of the National Academy of Sciences, USA*, **84**, 99–103.

Gurley, W.B., Czarnecka, E., Nagao, R.T. & Key, J.L. (1986). Upstream sequences required for efficient expression of a soybean heat shock gene. *Molecular and Cellular Biology*, **6**, 559–65.

Guy, C.L., Niemi, K.J. & Brambl, R. (1985). Altered gene expression during cold acclimation of spinach. *Proceedings of the National Academy of Sciences, USA*, **82**, 3673–7.

Hageman, R.H. & Flesher, D. (1960). The effect of anaerobic environment on the activity of alcohol dehydrogenase and other enzymes of corn seedlings. *Archives of Biochemistry and Biophysics*, **87**, 203–9.

Hahlbrock, K. (1981). Flavonoids. In *The Biochemistry of Plants*, Vol. 7, ed. E.E. Conn, pp. 425–9. New York: Academic Press.

Hahlbrock, K., Chappell, J., Jahnen, W. & Walter, M. (1985). In *Molecular Form and Function of the Plant Genome*, ed. L. van Vloten-Doting, G.S.P. Groot and T.C. Hall, pp. 129–40. New York: Plenum.

Hake, S., Kelley, P.M., Taylor, W.C. & Freeling, M. (1985). Coordinate induction of alcohol dehydrogenase 1, aldolase, and other anaerobic RNAs in maize. *Journal of Biological Chemistry*, **260**, 5050–4.

Hanson, A.D. & Hitz, W.D. (1982). Metabolic responses of mesophytes to plant water deficits. *Annual Review of Plant Physiology*, **33**, 163–203.

Hanson, A.D. & Jacobsen, J.V. (1984). Control of lactate dehydrogenase, lactate glycolysis and $\alpha$-amylase of $O_2$ deficit in barley aleurone layers. *Plant Physiology*, **75**, 566–72.

Hanson, A.D., Jacobsen, J.V. & Zwar, J.A. (1984). Regulated expression of three alcohol dehydrogenase genes in barley aleurone layers. *Plant Physiology*, **75**, 573–81.

Herrero, M.P. & Johnson, R.R. (1980). High temperature stress and pollen viability of maize. *Crop Science*, **20**, 796–800.

Higgins, C.F., Dorman, C.J., Stirling, D.A., Waddell, L., Booth, I.R., May, G. & Bremer, E. (1988). A physiological role for DNA supercoiling in the osmotic regulation of gene expression in *S. typhimurium* and *E. coli. Cell*, **52**, 569–84.

Hirschfield, I.N., Bloch, P.L., Van Bogelen, R.A. & Neidhardt, F.C. (1981). Multiple forms of lysyl-transfer ribonucleic acid synthetase in *Escherichia coli. Journal of Bacteriology*, **146**, 345–51.

Ho, T.-H.D., Nolan, R.C., Lin, L.-S., Brodl, M.R. & Brown, P.H. (1987). Regulation of gene expression in barley aleurone layers. In *Molecular Biology of Plant Growth Control*, ed. M. Jacobs and J.E. Fox, pp. 35–49. New York: Alan R. Liss.

Hooft van Huijsduijnen, R.A.M., Van Loon, L.C. & Bol, J.F. (1986). cDNA cloning of six mRNAs induced by TMV infection of tobacco and a characterization of their translation products. *EMBO Journal*, **5**, 2057–61.

Hsiao, T.C. (1973). Plant responses to water stress. *Annual Review of Plant Physiology*, **24**, 519–70.

Iida, H.I. & Yahara, I. (1985). Yeast heat-shock protein of Mr 48,000 is an isoprotein of enolase. *Nature*, **315**, 688–90.

Jackson, P.J., Naranjo, C.M., McClure, P.R. & Roth, E.J. (1985). The molecular response of cadmium resistant *Datura innoxia* cells to heavy metal stress. In *Cellular and Molecular Biology of Plant Stress*, ed. J.L. Key and T. Kosuge, pp. 145–60. New York: Alan R. Liss.

Jacobsen, J.V., Hanson, A.D. & Chandler, P.C. (1986). Water stress enhances expression of an α-amylase gene in barley leaves. *Plant Physiology*, **80**, 350–9.

Kaiser, W.M., Schröppel-Meier, G. & Wirth, E. (1986). Enzyme activities in an artificial stroma medium: An experimental model for studying effects of dehydration on photosynthesis. *Planta*, **167**, 292–9.

Kelley, P.M. & Freeling, M. (1982). A preliminary comparison of maize anaerobic and heat-shock proteins. In *Heat Shock From Bacteria to Man*, ed. M.J. Schlesinger, M. Ashburner and A. Tissieres, pp. 315–19. Cold Spring Harbor Laboratory Press.

Kelley, P.M. & Freeling, M. (1984a). Anaerobic expression of maize glucose phosphate isomerase I. *Journal of Biological Chemistry*, **259**, 673–7.

Kelley, P.M. & Freeling, M. (1984b). Anaerobic expression of maize fructose-1,6-diphosphate aldolase. *Journal of Biological Chemistry*, **259**, 14180–3.

Key, J.L., Lin, C.Y. & Chen, Y.M. (1981). Heat shock proteins of higher plants. *Proceedings of the National Academy of Sciences, USA*, **78**, 3526–30.

Kimpel, J.A. & Key, J.L. (1985). Presence of heat shock mRNAs in field grown soybeans. *Plant Physiology*, **79**, 672–8.

King, G.J. Hussey, C.E., Jr & Turner, V.A. (1986). A protein induced by NaCl in suspension cultures of *Nicotiana tabacum* accumulates in whole plant roots. *Plant Molecular Biology*, **7**, 441–9.

Kloppstech, K., Meyer, G., Schuster, G. & Ohad, I. (1985). Synthesis, transport and localization of nuclear coded 22-kd heat-shock protein in the chloroplast membranes of peas and *Chlamydomonas. EMBO Journal*, **4**, 1901–9.

Kuhn, D.N., Chappell, J., Boudet, A. & Hahlbrock, K. (1984). Induction of phenylalanine ammonia-lyase and 4-coumarate:CoA ligase mRNAs in cultured plant cells by UV light or fungal elicitor. *Proceedings of the National Academy of Sciences, USA*, **81**, 1102–6.

Laszlo, A. & St Lawrence, P. (1983). Parallel induction and synthesis of PDC and ADH in anoxic maize roots. *Molecular and General Genetics*, **192**, 110–17.

Lawton, M.A., Dixon, R.A., Rowell, P.M., Bailey, J.A. & Lamb, C.J. (1983). Rapid induction of the synthesis of phenylalanine ammonia-lyase and of chalcone synthase in elicitor-treated plant cells. *European Journal of Biochemistry*, **129**, 593–601.

Levitt, J. (1980). *Response of Plants to Environmental Stress*, Vol. I. New York: Academic Press.

Lin, C.Y. & Key, J.L. (1967). Dissociation and reassembly of polyribosomes in relation to protein synthesis in the soybean root. *Journal of Molecular Biology*, **26**, 237–47.

Lin, C.Y., Roberts, J.K. & Key, J.L. (1984). Acquisition of thermotolerance in soybean seedlings: synthesis and accumulation of heat shock proteins and their cellular localization. *Plant Physiology*, **74**, 152–60.

Lindquist, S. (1981). Regulation of protein synthesis during heat shock. *Nature*, **294**, 311–14.

Llewellyn, D.J., Finnegan, E.J., Ellis, J.G., Dennis, E.S. & Peacock, W.J. (1987). Structure and expression of an alcohol dehydrogenase 1 gene from *Pisum sativum* (cv. 'Greenfeast'). *Journal of Molecular Biology*, **195**, 115–23.

Mayoral, M.L., Atsmon, D., Gromet-Elhanan, Z. & Shimshi, D. (1981). Effect of water stress on enzyme activities in wheat and related wild species: Carboxylase activity, electron transport and photophosphorylation in isolated chloroplasts. *Australian Journal of Plant Physiology*, **8**, 385–94.

Nagao, R.T., Czarnecka, E., Gurley, W.B., Schöffl, F. & Key, J.L. (1985). Genes for low-molecular-weight heat shock proteins of soybeans: Sequence analysis of a multigene family. *Molecular and Cellular Biology*, **5**, 3417–28.

Nagata, K., Saga, S. & Yamada, K.M. (1986). A major collagen-binding protein of chick embryo fibroblasts is a novel heat shock protein. *Journal of Cell Biology*, **103**, 223–9.

Nieto-Sotelo, J. & Ho, T.-H.D. (1987). Absence of heat shock protein synthesis in isolated mitochondria and plastids from maize. *Journal of Biological Chemistry*, **262**, 12288–92.

Nover, L., Scharf, K.D. & Neumann, D. (1983). Formation of cytoplasmic heat shock granules in tomato cell cultures and leaves. *Molecular and Cellular Biology*, **3**, 1648–55.

Okimoto, R., Sachs, M.M., Porter, E.K. & Freeling, M. (1980). Patterns of polypeptide synthesis in various maize organs under anaerobiosis. *Planta*, **150**, 89–94.

Ostrem, J.A., Olson, S.W., Schmitt, J.M. & Bohnert, H. (1987). Salt stress increases the level of translatable mRNA for phosphoenolpyruvate carboxylase in *Mesembryanthemum crystallinum*. *Plant Physiology*, **84**, 1270–5.

Pelham, H.R.B. & Bienz, M. (1982). A synthetic heat-shock promoter element confers heat-inducibility on the herpes simplex virus thymidine kinase gene. *EMBO Journal*, **1**, 1473–7.

Philips, T.A., Van Bogelen, R.A. & Neidhardt, F.C. (1984). *lon* gene product of *Escherichia coli* is a heat-shock protein. *Journal of Bacteriology*, **159**, 283–7.

Rauser, W.E. (1984). Isolation and partial purification of cadmium-binding protein from roots of the grass *Agrostis gigantea*. *Plant Physiology*, **74**, 1025–9.

Reese, R.N. & Wagner, G.J. (1987). Properties of tobacco (*Nicotiana tabacum*) cadmium-binding peptide(s). *Biochemistry Journal*, **241**, 641–7.

Roberts, J.K.M., Andrade, F.H. & Anderson, I.C. (1985). Further evidence that cytoplasmic acidosis is a determinant of flooding intolerance in plants. *Plant Physiology*, **77**, 492–4.

Roberts, J.K.M., Callis, J., Wemmer, D., Walbot, V. & Jardetzky, O. (1984). Mechanisms of cytoplasmic pH regulation in hypoxic maize root tips and its role in survival under hypoxia. *Proceedings of the National Academy of Sciences, USA*, **81**, 3379–83.

Rochester, D.E., Winter, J.A. & Shah, D.M. (1986). The structure and expression of maize genes encoding the major heat shock protein, hsp70. *EMBO Journal*, **5**, 451–8.

Rowland, L.J. & Strommer, J.N. (1985). Insertion of an unstable element in an intervening sequence of maize *Adh1* affects transcription but not processing. *Proceedings of the National Academy of Sciences, USA*, **82**, 2875–9.

Ryan, C.A., Bishop, P.D., Walker-Simmons, M., Brown, W. & Graham, J.S. (1985). The role of pectic fragment of the plant cell wall in the response to biological stresses. In *Cellular and Molecular Biology of Plant Stress*, ed. J.L. Key and T. Kosuge, pp. 319–34. New York: Alan R. Liss.

Sachs, M.M., Dennis, E.S., Gerlach, W.L. & Peacock, W.J. (1986). Two alleles of maize *alcohol dehydrogenase 1* have 3′ structural and poly(A) addition polymorphisms. *Genetics*, **113**, 449–67.

Sachs, M.M. & Freeling, M. (1978). Selective synthesis of alcohol dehydrogenase during anaerobic treatment of maize. *Molecular and General Genetics*, **161**, 111–15.

Sachs, M.M., Freeling, M. & Okimoto, R. (1980). The anaerobic proteins of maize. *Cell*, **20**, 761–7.

Sachs, M.M. & Ho, T.-H.D. (1986). Alteration of gene expression during environmental stress in plants. *Annual Review of Plant Physiology*, **37**, 363–76.

Schlesinger, M.J., Ashburner, M. & Tissieres, A. (1982). *Heat Shock From Bacteria to Man*. Cold Spring Harbor Laboratory Press.

Schöffl, F. & Key, J.L. (1982). An analysis of mRNA's for a group of heat shock proteins of soybean using cloned cDNA's. *Journal of Molecular and Applied Genetics*, **1**, 301–14.

Schöffl, F., Raschke, E. & Nagao, R.T. (1984). The DNA sequence analysis of soybean heat-shock genes and identification of possible regulatory promoter elements. *EMBO Journal*, **3**, 2491–7.

Schönfelder, M., Horsch, A. & Schmid, H.-P. (1985). Heat shock increases the synthesis of the poly(A)-binding protein in HeLa cells. *Proceedings of the National Academy of Sciences, USA*, **82**, 6884–8.

Schuster, G., Even, D., Kloppstech, K. & Ohad, I. (1988). Evidence for protection by heat-shock proteins against photoinhibition during heat-shock. *EMBO Journal*, **7**, 1–6.

Schwartz, D. (1969). An example of gene fixation resulting from selective advantage in suboptimal conditions. *American Naturalist*, **103**, 479–81.

Showalter, A.M., Bell, J.N., Cramer, C.L., Bailey, J.A., Varner, J.E. & Lamb, C.J.

(1985). Accumulation of hydroxyproline-rich glycoprotein mRNAs in response to fungal elicitor and infection. *Proceedings of the National Academy of Sciences, USA*, **82**, 6551–5.

Singh, N.K., Handa, A.K., Hasegawa, P.M. & Bressan, R.A. (1985). Proteins associated with adaptation of cultured tobacco cells to NaCl. *Plant Physiology*, **79**, 126–37.

Singh, N.K., La Rosa, C., Handa, A.K., Hasegawa, P.M. & Bressan, R.A. (1987). Hormonal regulation of protein synthesis associated with salt tolerance in plant cells. *Proceedings of the National Academy of Sciences, USA*, **84**, 739–43.

Somssich, I.E., Schmeizer, E., Bollmann, J. & Hahlbrook, K. (1986). Accumulation of hydroxyproline-rich glycoprotein mRNAs in response to fungal elicitor and infection. *Proceedings of the National Academy of Sciences, USA*, **82**, 6551–5.

Storti, R.V., Scott, M.P., Rich, A. & Pardue, M.L. (1980). Translational control of protein synthesis in response to heat shock in *D. melanogaster* cells. *Cell*, **22**, 825–34.

Ungewickell, E. (1985). The 70-kd mammalian heat shock proteins are structurally and functionally related to the uncoating protein that releases clathrin triskelia from coated vesicles. *EMBO Journal*, **4**, 3385–91.

Van Loon, L.C. (1985). Pathogenesis-related proteins. *Plant Molecular Biology*, **4**, 111–16.

Vierling, E. & Key, J.L. (1985). Ribulose 1,5-bisphosphate carboxylase synthesis during heat shock. *Plant Physiology*, **78**, 155–62.

Vierling, E., Mishkind, M.L., Schmidt, G.W. & Key, J.L. (1986). Specific heat shock proteins are transported into chloroplasts. *Proceedings of the National Academy of Sciences, USA*, **83**, 361–5.

Walker, J.C., Howard, E.A., Dennis, E.S. & Peacock, W.J. (1987). DNA sequences required for anaerobic expression of the maize *Alcohol dehydrogenase-1* gene. *Proceedings of the National Academy of Sciences, USA*, **84**, 6624–8.

Webb, M. (1979). The metallothioneins. In *The Chemistry, Biochemistry, and Biology of Cadmium*, ed. M. Webb, pp. 195–266. Amsterdam: Elsevier/North Holland.

Weiderrecht, G., Shuey, J.D., Kibbe, W.A. & Parker, C.S. (1987). The Saccharomyces and Drosophila heat shock transcription factors are identical in size and DNA binding properties. *Cell*, **48**, 507–15.

Weiser, C.J. (1970). Cold resistance and injury in woody plants. *Science*, **169**, 1269–78.

Winter, K. (1974). Enfluß von Wasserstreb auf die Aktivität der Phosphoenol-pyruvat-Carboxylase bei *Mesembryanthemum crystallinum* (L.). *Planta*, **121** 147–53.

Wu, C., Wilson, S., Walker, B., Dawid, I., Paisy, T., Zimarino, V. & Ueda, H. (1987). Purification and properties of *Drosophila* heat shock activator protein. *Science*, **238**, 1247–53.

Xiao, C.-M. & Mascarenhas, J.P. (1985). High temperature-induced thermotolerance in pollen tubes of *Tradescantia* and heat-shock protein. *Plant Physiology*, **78**, 887–9.

Younis, H.M., Boyer, J.S. & Govindjee (1979). Conformation and activity of chloroplast coupling factor exposed to low chemical potential of water in cells. *Biochimica et Biophysica Acta*, **548**, 328–40.

Younis, H.M., Weber, G. & Boyer, J.S. (1983). Activity and conformational changes in chloroplast coupling factor induced by ion binding: Formation of a magnesium–enzyme–phosphate complex. *Biochemistry*, **22**, 2505–12.

# 10 Plant tissue and protoplast culture: applications to stress physiology and biochemistry

## Introduction

Salinity and drought, two environmental stresses frequently found together, are major barriers to productivity of agricultural crops throughout the world. Crops exposed to these stressful environments are observed initially to have reduced growth rates. If the stress is more severe the response is manifested visually in a number of specific and recognisable symptoms, many of which are common to both salinity and drought. However, specific ion toxicity responses (e.g. marginal leaf burn) are observed in plants exposed to excess salinity. Since salinity and drought have many common responses some of the information presented in this review will be combined under the general umbrella of environmental stress.

This chapter discusses the use of tissue and protoplast culture as a means to understand better the cellular processes related to stress tolerance with the expectation that these techniques will provide alternative methods for screening germplasm and assist in identifying useful material for incorporation into crop improvement programmes. Of equal importance is the ability to compare physiological and biochemical processes of cells and protoplasts selected for stress tolerance against unselected (wild-type) lines and to relate this to genetic regulation at the molecular level.

The techniques of tissue and protoplast culture are important for application of molecular biology to genetic manipulation of plants. The field of plant pest–host interactions has been significantly advanced by the use of protoplasts and subsequent tissue culture and regeneration of pest-resistant plants (Harrison & Mayo, 1983). Somaclonal variation has provided important genetic material both for genetic studies and for selection of desired traits in plants (Scowcroft, Larkin & Brettell, 1983; Maddock & Semple, 1986). More recently protoplast fusion has made available plant cell systems (cybrids) which greatly advance the study of organelle function and genetics while providing much enhanced genetic variability (Kumar & Cocking, 1987). Some examples of the application of cell and tissue culture systems in breeding for stress tolerance are outlined by Yeo & Flowers (Chapter 12).

The preceding discussion has considered, in general, the potential advantages of the use of tissue and protoplast culture. Before detailing information on the use of these techniques in understanding the effects of stress on plants it is important to present some of the limitations. Cells and protoplasts used in physiological and biochemical studies will provide information on cellular and molecular processes; however, mechanisms related to stress frequently depend on the functions and structures of an intact plant (Passioura, 1986; Rains, Croughton & Croughan, 1986). For example in cell culture systems the cell-to-cell interaction of fairly rigid plant structures is lost and the role of cell turgor and other physical mechanisms in the response of plants to stress cannot be evaluated. If a trait is defined at the cellular level it is essential that the impact of this trait be characterised at the whole-plant level and that the character be shown to be inherited and transferred to succeeding generations. This requires regeneration of cells to plants. The list of regenerated plant species from tissue culture is quite long and continues to grow (Rains *et al.*, 1986). However, the list is considerably shorter when protoplasts are the starting material for regeneration. One would anticipate that the ability to regenerate protoplasts to plants will continue to improve and the list will match the list of cells regenerated to plants. Other limitations of tissue and protoplast culture include the following. (1) Culture media are very different in composition from those in which plant roots normally grow. Cultured cells are provided with high concentrations of sucrose (up to 90 mM) and inorganic ions (up to 100 mM). This is of importance to studies on osmotic adjustment and there has been at least one recent attempt to use a minimal medium for experiments with cultured cells. (2) Analyses of ion contents have generally been confounded with inadequate washing procedures and the fact that the apparent free space of cultured cells is some 50% of their volume (Dracup *et al.*, 1986).

### Cellular responses to water stress

*In vitro* plant culture systems offer a number of advantages when studying the response of plant cells to environmental stress (Hasegawa, Bressan & Handa, 1987; Stavarek & Rains, 1984*a*). Some of these advantages are more applicable to one stress than to others. Hasegawa *et al.* (1987) have discussed these unique applications in investigations on water stress. Some of the advantages are: (i) cells grown in liquid suspension cultures are exposed to a uniform water potential which can be altered similarly for the entire population of cells; (ii) cells growing in culture are uniform in development so that any biochemical or physiological changes related to water stress will represent changes in growing cells (in contrast the intact plant system is a heterogeneous population of cells with proportionately few of these cells actively growing); (iii) the relationship between water potential, solute potential and pressure potential can be evaluated independent of the confounding factors of cell-to-cell interaction. However, the interaction is undoubtedly important to intact plants adjusting to stress.

Osmotica used to induce water deficits in cultured cells can be of three types: (i) freely permeating materials, such as glycerol, which freely penetrate both the cell wall and the plasmalemma; (ii) absorbable osmotica, such as mannitol, for which the plasmalemma is relatively impermeable but which readily penetrate the cell wall; (iii) non-absorbable osmotica that are too large even to penetrate the cell wall. Their effects on cell water relations are different: freely permeating osmotica affect the water potential of the cells, but have no effect on cell turgor or volume; absorbable osmotica can cause intracellular water loss to the extent that the plasmalemma pulls away from the cell wall leading to plasmolysis; in contrast, the use of non-absorbed osmotic agents, such as high molecular weight polyethylene glycol (PEG), results in a collapse of the cell (cytorrhysis) with no observed plasmolysis (Carpita *et al.*, 1979).

### Growth

Cells selected for tolerance to water stress have been demonstrated in a number of plant species (Handa *et al.*, 1983; Heyser & Nabors, 1981; Stavarek & Rains, 1984*b*). High molecular weight PEG is commonly used to induce water deficits (Hasegawa *et al.*, 1984). This provides an opportunity to evaluate intracellular osmotic adjustment of plants exposed to water stress. If information is desired on the effect of specific osmotic substances on cellular adjustment to osmotic stress PEG can be used as an independent osmotic regulator and/or in combination with an absorbable osmoticum. In these situations the interaction of osmoticum produced by the cell with the osmoticum absorbed by the cell can be evaluated.

Growth characteristics of cells exposed to water stress mimic some of the structural responses of organised plant tissues. A frequently observed response of plants exposed to water stress is a reduction in cell size (Cutler, Rains & Loomis, 1977). This cellular phenomenon was observed in tomato cells stressed with PEG (Handa *et al.*, 1983). Concomitantly with a decrease in cell size with increasing osmotic stress was a reduction in fresh weight. In contrast the dry weight was not affected.

Cytorrhysis, compared to plasmolysis, creates different relationships between the plasmalemma, cytosol and cell wall possibly resulting in an alteration in membrane ion transport. Evidence is available showing the interaction between the osmotic agent used to stress cells and the transport characteristics. Reuveni *et al.* (1987) demonstrated that when carrot cells grown in suspension culture were exposed to sorbitol, turgor was decreased to zero and the membrane was depolarised. The level of ATP decreased and protons were extruded into the medium causing acidification. When NaCl was used as the osmoticum the decrease in turgor happened in the absence of changes in intracellular ATP. Their conclusions were that sorbitol is only slowly absorbed and this results in a triggering of a proton extrusion pump that utilises ATP. In the presence of a rapidly permeating substance such as NaCl the cells utilise a $Na^+/H^+$ antiporter and no ATP is required. This could have implications for the relative metabolic costs of cells exposed to different osmotic substances. A similar NaCl-induced $Na^+/H^+$ antiport activity has been observed in suspension cultures of sugar beet (Blumwald & Poole, 1987).

The common response of both cultured cells and cells comprising the body of a plant when these two systems are exposed to water stress is the requirement for osmotic adjustment. It seems reasonable then to expect that information obtained at the cellular level should enhance our understanding of the biochemical and physiological response of plants exposed to water stress.

### Osmotic adjustment – inorganic

Cells tolerant to water stress have elaborated mechanisms to maintain a favourable osmotic gradient. A generally observed response is absorption of osmoticum from the external environment. Inorganic ions such as Na, K and Cl are frequently found in excess in stressful environments and may increase in cultured cells exposed to water stress, though the contribution of the individual ions may not be proportional to the external concentration of those ions. It is notable that ions are not greatly used by plants (as opposed to cells) for osmotic adjustment to water stress, other than by concentration of the normal vacuolar solutes by loss of water (see Flowers & Yeo, 1986). K was found to accumulate in tomato cells when exposed to water stress, although reducing sugars provided a larger amount of internal osmoticum than did K. Similar to many osmoregulating systems there was no relationship between turgor and growth. Tobacco cells selected for tolerance to NaCl and then exposed to PEG have been compared with non-selected lines. The selected lines were more tolerant and responded by developing a more negative intracellular osmotic pressure due to dehydration, uptake of inorganic ions from the external environment and by producing organic osmoticum (Heyser & Nabors, 1981). Conversely, carrot cells adapted to mannitol were found to have an overall greater tolerance to salinity (Harms & Oertli, 1985). NaCl-tolerant citrus cells were found to be more tolerant to PEG induced stress than were cells from a non-selected line (Ben-Hayyim, 1987). In the Shamouti orange cell lines selected for salt tolerance, exposure to PEG resulted in an increase in internal K concentrations independent of the external Na concentrations. The conclusion was that K plays a key role in the growth of plant cells exposed to stress (Ben-Hayyim, Kalkafi & Ganmore-Neuman, 1987): a role for K in plants exposed to environmental stress has been discussed in a number of reviews (Rains, 1972; Leigh & Wyn Jones, 1984). The observation that selection for tolerance to a particular stress conveys tolerance to another stress lends support to the concept that certain environmental stresses, e.g. heat, cold, salt and water, have common physico-chemical processes responsible for the stress and therefore selection for one stress may provide biological material tolerant to another stress (Steponkus, 1980; see also Chapter 1).

*Osmotic adjustment – organic*

Osmotically active substances have been reported to increase in plant cells when these cells are exposed to water stress. The substances include both inorganic and organic ions. Accumulation of organic substances in response to stress may cause toxicity although the potential toxicity of organic substances is frequently less than iso-osmotic levels of inorganic ions. A frequent observation is that organic substances, e.g. proline and glycine betaine, protect proteins against dehydration (Paleg, Stewart & Bradbeer, 1984) and the protection is proportional to the concentration of these two substances. Since proline and glycine betaine are osmotically active they perform a dual function – protect against dehydration of proteins and maintain a more favourable osmotic gradient in the cell.

Tomato cells exposed to PEG to induce water deficits responded with an increase in dry weight/fresh weight ratio, a loss in turgor and rapid accumulation of proline (Handa *et al.*, 1986). Proline levels continued to increase as turgor recovered during osmotic adjustment. These results suggest that accumulation of proline depended not only on cell water potential and/or initial loss of turgor, but more closely responded to cell solute potential. External additions of proline enhanced tolerance to water stress. Duncan & Widholm (1987) observed proline accumulation by maize callus when water stress was induced by mannitol. They correlated proline accumulation with increased tolerance to cold and found that any condition enhancing proline concentration in the cells, water stress, ABA or proline in the culture media, also enhanced cold tolerance. These observations suggest again that when plant cells are exposed to environmental stresses responses to those stresses may involve similar tolerance processes or mechanisms.

A more comprehensive review of the significance of these organic osmotica in the tolerance of plant cells to water stress can be found in a paper by Hasegawa *et al.* (1987). The significance of organic osmotica in tolerance processes of intact plant systems is discussed in Chapters 7 and 8.

### Cellular response to salinity

Cells exposed to saline stress encounter reduced water availability, ion toxicity and reduced availability of essential nutrients. These cellular level responses are also reflected at the whole-plant level. An understanding of these cellular responses will undoubtedly contribute to an understanding of the response of a plant growing in a saline environment.

### *Osmotic adjustment – inorganic*

Cells exposed to excessive levels of salinity have to acquire essential nutrients from a milieu with a preponderance of ions that are potentially toxic and non-essential. In this ionic environment the success of a plant cell will require intracellular tolerance and/or specific acquisition of nutrients essential for normal metabolic functioning. The cell is also exposed to an unfavourable water balance with an absolute requirement to maintain an internal osmotic regulation that favours uptake of water into the cell (Stavarek & Rains, 1984*b*).

A number of investigators have successfully selected cell lines which have higher tolerance to salinity than the line from which they were selected (see Spiegel-Roy & Ben Hayyim, 1985; Rains *et al.*, 1986 for a list of plant species). An evaluation of these selected lines demonstrates a number of differences in ionic status and cell wall regulation, but there are relatively few cases where the salinity tolerance of whole plants that have been regenerated from this material have been determined (see Yeo & Flowers, Chapter 12).

Ben-Hayyim *et al.* (1987) demonstrated that cultured citrus cells selected for tolerance to NaCl were most tolerant to polyethylene glycol, followed by NaCl and then $CaCl_2$. The exposure of the cells to any of these osmotic agents resulted in an increase in intracellular K. The authors concluded that K played a key role in the growth of cells exposed to salt. Other researchers have also suggested that K may play a significant role in the response of plant cells to salinity (Rains, 1972; Croughan, Stavarek & Rains, 1978;

Watad, Rheinhold & Lerner, 1983; Leigh & Wyn Jones, 1984; Rains *et al.*, 1986; Binzel *et al.*, 1987; Bourgeais, Guerrier & Strullu, 1987).

Plant cells selected for tolerance to stress show varied responses to the imposed osmotic gradients. In adapted cells, tolerance to salinity or to water stress was not found to increase proportionately with increases in turgor (Handa *et al.*, 1983; Binzel *et al.*, 1985). It was suggested from these observations and from studies by Heyser & Nabors (1981) that no relationship existed between turgor and growth and that stress adaptation may alter the relationship between turgor and cell expansion (see also Chapter 6).

Alteration in hormone levels have been observed in plants exposed to environmental stress (Davies *et al.*, 1986; Chapter 5). The role played by these hormones in regulating growth of stressed plants may not be direct, though turgor is known to be influenced by hormones and in environmentally stressed plants turgor maintenance is a critical function (Cleland, 1986; Jones *et al.*, 1987). Abscisic acid (ABA) has been suggested to affect cell wall extensibility and thus turgor. La Rosa and co-workers (1987) found that ABA stimulated osmotic adjustment of tobacco cells during adaptation to NaCl. Although inorganic ions did not account for the enhanced osmotic adjustment stimulated by ABA, uptake of inorganic ions was increased thus maintaining the ion pool during ABA-induced growth. Evaluation of intracellular contents showed that the increased osmotic adjustment was due primarily to higher levels of sucrose, reducing sugars and proline. No specific role for ABA was defined by the authors, but they speculated that ABA may alter the compartmentation of osmoticum with organic solutes concentrated in the cytoplasm and inorganic solutes in the vacuole.

A more general role of ABA in stress tolerance has been found in carrot cells. When a suspension culture of carrot cells was exposed to ABA and then selected for tolerance to freezing, the ABA-treated cells were found to be more tolerant to the stress (Reaney & Gusta, 1987). These results provide further evidence for the presence of common mechanisms conveying tolerance to many of the environmental stresses.

*Osmotic adjustment – organic*

Osmotic adjustment by plant cells in response to an increasing saline environment can be mediated by an alteration in intracellular concentrations of both inorganic and organic ions (Wyn Jones, 1980, 1984; Aspinall, 1986; Flowers & Yeo, 1986; Grumet & Hanson, 1986; Moftah & Michel, 1987).

The organic solutes shown to be related to stress adaptation include sugars, polyols, tertiary N compounds, amino acids and organic acids. Of particular note are compounds such as proline, glycine betaine and reducing sugars (Jefferies, 1980; Stewart & Lahrer, 1980; Hanson & Hitz, 1982; Wyn Jones & Gorham, 1983).

A number of chapters in this volume (especially Chapters 5 and 6) provide a more thorough discussion of osmotic adjustment by intact plants and tissues in response to environmental stress and the role of osmotically active solutes in this response. The following section focuses on the role of organic osmotica in the response of plant cells to salt stress. Cultured plant cells offer the opportunity to evaluate the effect of both internally synthesised and externally administered organic osmotica.

A frequently observed response of plant cells exposed to saline stress is the accumulation of proline. Two cell lines of tobacco, one resistant and the other sensitive to growth inhibition by NaCl, accumulated proline when exposed to 1.5% w/v NaCl in the growth media (Dix & Pearce, 1981). The NaCl sensitive line accumulated proline more rapidly than did the resistant line, though the levels accumulated were not adequate to provide osmotic protection against salt stress. The authors suggested that proline accumulation may have a protective role other than osmoregulation and may be symptomatic of stress injury, the nature of which was not discussed.

*Brassica napus* cells have been selected for tolerance to $Na_2SO_4$ (Chandler & Thorpe, 1987). When selected cells were compared with non-selected cells in response to $Na_2SO_4$ salinity the selected cells grew better and showed less negative cell water potential than the non-selected cells. Both cell lines showed osmotic adjustment and proline accumulation. However, proline accumulation was related to inhibition of growth and did not play a significant role in osmotic adjustment.

Cowpea cell lines selected for tolerance to NaCl were found to accumulate proline when exposed to NaCl salinity while the salt-sensitive line did not accumulate this amino acid (Pandey & Ganapathy, 1985). When proline was added to the culture medium containing 100 mM NaCl the fresh weight, dry weight and proline content of both cell lines increased. When KCl salinity was substituted for NaCl salinity the growth and free proline levels of the resistant line were reduced: not so the sensitive line. These authors, in contrast to Dix & Pearce, concluded that accumulation of proline and biomass, simultaneously, provides evidence that proline plays a specific role in reducing salt stress and that accumulation of proline by plant cells in response to salinity should not be considered as only symptomatic of stress injury. The function of proline and other organic molecules in mediating stress responses remains, as the title of their paper states, an enigma.

The specific role of organic osmoticum in cultured plant cells remains unclear. Recent results, however, have provided some evidence as to their possible function. Cultured barley embryos were treated with exogenous proline and glycine betaine (Lone *et al.*, 1987). The embryos were germinated and shoot elongation was evaluated under saline conditions. The presence of proline in the embryo culture media appeared to enhance $K^+/Na^+$ discrimination in the transport from the root to the shoot and assisted in retaining salts in the root, thus excluding them from the shoot. The authors suggested that the effect of proline was to stabilise membranes and protect enzymes from the deleterious effects of salt thus allowing growth to proceed. Proline did not seem to be involved with overall osmotic adjustment. These experiments of Lone and co-workers demonstrate the potential of combining culture techniques and intact plant tissue experiments. These organic solutes function as protein stabilisers and mediate osmotic adjustment. The extent of these functions depends on the compartmentation of the solutes and the size of the osmotic volume involved.

### Metabolic responses: proteins

One of the metabolic responses of plants exposed to environmental stress is the production of proteins which may be qualitatively and/or quantitatively different from those produced in the absence of the stress (see Chapter 9 for general discussion). In some cases these responses have been found to depend on genotype: for example, when a salt-tolerant cultivar and a salt-sensitive cultivar of barley were exposed to salt stress the shoot tissue responded by synthesising proteins which were cultivar specific. Five new proteins not found in the salt-sensitive barley were identified in the salt-tolerant cultivar (Ramagopal, 1987). No differences in proteins were found in the roots of either cultivar.

Cell lines exposed to salt stress also respond by synthesising specific proteins. Tobacco cells adapted to grow on NaCl produced new polypeptides or increased the existing polypeptide bonds when exposed to NaCl or PEG (Singh *et al.*, 1985). During adaptation of these tobacco cells to stress an increased synthesis of a 26 kDa polypeptide was observed. This increase correlated with the period of osmotic adjustment and growth of the cultured cells. The 26 kDa polypeptide induced during adaptation to salt stress accounted for 10% of the total cellular protein. The authors proposed an involvement of this protein in adaptation to salt and osmotic stress, although they did not identify a specific physiological or biochemical function. Koch, Wells & Grisebach (1987) osmotically stressed soybean calli and observed an increase in NADPH-dependent flavone synthetase. Although a specific enzymatic function was observed there was no obvious relationship between the enzyme activity with tolerance to stress.

Singh and co-workers (1987*b*) found that tobacco cells selected to tolerate 500 mM NaCl when exposed to salt stress responded by synthesising more protein and endogenous abscisic acid. When ABA was supplied exogenously a 26 kDa protein was synthesised, though NaCl stress was required for the induced protein to accumulate. There has been a suggestion that one role of ABA in stress responses is through alteration in cell wall elasticity (Chapter 6; Cleland, 1986). Singh and co-workers (1987b) concluded from two lines of evidence that ABA does not alter cell wall elasticity. First, cell wall elasticity did not affect osmotic adjustment which occurs as plasmolysed cells are regaining volume and secondly, ABA-treated cells show osmotic adjustment occurring as turgor is regained and these cells are growing faster than untreated cells.

The 26 kDa protein synthesised by salt-adapted tobacco cells has been further characterised (Singh et al., 1987a). The protein makes up approximately 12% of the total cellular protein and has been resolved into two forms. These two forms have been designated osmotin I and osmotin II and occur in a 2:3 ratio. The forms are distinct with osmotin I soluble in an aqueous phase and osmotin II soluble in detergent. The proteins accumulate as inclusion bodies in the vacuole and are only sparsely distributed in the cytoplasm.

The specific role(s) of proteins induced by salt stress is unclear (Chapter 9). Singh and co-workers (1987a) speculated that these proteins could serve as storage proteins which accumulate when cells are stressed and growth is restricted. Cross-reaction studies with many common storage proteins did not support this speculation. Another possibility is the involvement of these proteins in osmotic adjustment through facilitation of accumulation or compartmentation of osmotic materials or by affecting metabolic and structural changes. The understanding of the role of these proteins will certainly assist our understanding of the biochemical and molecular basis of salt tolerance in plants.

### Use of protoplasts in studying stress

The use of protoplasts in studies of stress physiology and biochemistry expands the advantages of cell culture systems discussed in the preceding sections. Additional applications are related to the fusion of protoplasts. Intraspecific and interspecific protoplast fusion greatly enhance genetic variability of the fused protoplasts (Kumar & Cocking, 1987). The resulting somatic hybrids provide cells which can be used for selection of specific traits (e.g. environmental stress tolerance) provided by one or both donor cells and for basic studies on cytoplasmic and nuclear inheritance of desired characteristics.

Refinement of techniques involved in protoplast isolation has enhanced the success of isolation of vacuoles and other intracellular structures

(Wagner, 1983; Alibert *et al.*, 1985). These techniques are important for understanding the role of cell compartments in partitioning osmotica in stressed cells (Leigh, 1983). The success and application of this approach requires regeneration of protoplasts to plants. The number of plant species taken through protoplast culture and regenerated to plants continues to increase (Abdullah, Cocking & Thompson, 1986; Cocking, 1986; Imbrie-Milligan & Hodges, 1986).

## Stress tolerance

The literature is less extensive on the use of protoplasts in stress-tolerance investigations; however, some applications have been attempted. For example, in one study protoplasts were isolated from the leaves of a wild relative of tomato shown to be salt tolerant and from a salt-sensitive, cultivated species (Rosen & Tal, 1981). In the presence of NaCl the plating efficiency (number of surviving cells/number of cells applied to the plate) of the wild relative was greater than the cultivated, sensitive cultivar. Proline, when added to the culture media, was found to enhance the plating efficiency of the salt-sensitive cultivar but not the wild, salt-tolerant relative. These results suggest that traits related to salt tolerance are expressed by the isolated protoplasts and that the response of protoplasts to environmental stress can be manipulated, i.e. the proline response.

A more significant body of literature focuses on the use of protoplasts in understanding processes related to stress tolerance. The role of Ca in salt tolerance has been evaluated using maize root protoplasts. Exposure of the plasmalemma directly to external media revealed a non-specific replacement of Ca by salt. Sodium was found to replace Ca though this could be reversed by adding more Ca (Lynch, Cramer & Läuchli, 1987). This approach assists in understanding the role of specific ion interaction in enhancing salt tolerance and is potentially applicable to studies on the molecular basis for ion specificity of plant membranes.

Caution is necessary when using such techniques. For example, proto-plasts were isolated from the epidermis and mesophyll of *Commelina communis* and when suction was applied to the cells, the cell membrane potentials became more negative (Pantoja & Willmer, 1986). The cell membrane was stretched by the suction action, simulating changing cell turgor, which had a significant effect on ion fluxes. The authors suggested that caution is needed when interpreting ion flux data from studies on protoplasts.

Similar problems are apparent when using isolated cellular structures such as vacuoles. The use of vacuoles in furthering our understanding of the role of compartmentation of osmotica has to be tempered with the frequent observation that vacuoles are leaky and are variable in response (Leigh & Tomos, 1983).

Protoplasts offer a major advantage in that they are a cellular system devoid of the cell wall, a significant physical barrier. The absence of this physical barrier mediates the direct genetic manipulation such as insertion of specific genetic information (DNA) directly into receptor cells. Theoretically the desired genetic information can be directly introduced from a diverse source, expressed and sexually inherited. A number of examples of this technique have been demonstrated, including direct gene uptake by soybean protoplasts mediated by electrophoretic procedures (Lin, Odell & Schreiner, 1987). This approach to enhancing genetic variability will become increasingly more valuable in the study of genetic regulation of physiological and biochemical processes including those processes related to environmental stress tolerance.

### Conclusions

Genetic manipulation of plants as a means to improve performance has been and continues to be very successful. Modern plant breeding using natural genetic variability, and to a limited extent induced variability, revolutionised agriculture by providing cultivars with resistance to both biological and environmental stress and with desired quality characteristics while maintaining crop yield.

Cell and protoplast culture are tools that add to this successful effort in plant improvement. As the need for food production expands into marginal environments and as the demand for ecologically sound farming systems gives impetus to genetic solutions for management practices there will be an increasing need for genetically 'tailored' crops to fit these situations. The potential for genetic modification will be greatly enhanced by better understanding of processes related to stress tolerance.

Physiological and biochemical studies of selected cell lines have provided and will continue to provide significant information on the cellular processes and mechanisms associated with tolerance to stress. Comparative studies of selected and non-selected cell lines have identified a number of metabolic processes correlated with tolerance to environmental stress. An understanding of these processes could provide the plant breeder with markers to assist in selecting plants with desired traits. Concomitantly the characterisation of the physiological and biochemical mechanisms which regulate plant responses to stress will be essential for the application of molecular genetics to the improvement of plant germplasm.

### References

Abdullah, R., Cocking, E.C. & Thompson, J.A. (1986). Efficient plant regeneration from rice protoplasts through somatic embryogenesis. *Bio/Technology*, **4**, 1087–90.

Alibert, G., Bondet, A.M., Canut, H. & Rataboul, P. (1985). Protoplasts in studies of vacuolar storage compounds. In *The Physiological Properties of Plant Protoplasts*, ed. P.E. Pilte, pp. 105–15. Berlin: Springer.

Aspinall, D. (1986). Metabolic effects of water and salinity stress in relation to expansion of the leaf surface. *Australian Journal of Plant Physiology*, **13**, 59–73.

Ben-Hayyim, G. (1987). Relationship between salt tolerance and resistance to polyethylene glycol-induced water stress in cultured citrus cells. *Plant Physiology*, **85**, 430–4.

Ben-Hayyim, G., Kalkafi, U. & Ganmore-Neuman, R. (1987). Role of internal potassium in maintaining growth of cultured citrus cells on increasing NaCl and $CaCl_2$ concentrations. *Plant Physiology*, **85**, 434–40.

Binzel, M.L., Hasegawa, P.M., Handa, A.K. & Bressan, R.A. (1985). Adaptation of tobacco cells to NaCl. *Plant Physiology*, **79**, 118–25.

Binzel, M.L., Hasegawa, P.M., Rhodes, D., Handa, S., Handa, A.K. & Bressan, R. (1987). Solute accumulation in tobacco cells adapted to NaCl. *Plant Physiology*, **84**, 1408–15.

Blumwald, E. & Poole, R.J. (1987). Salt tolerance in suspension cultures of sugar beet. Induction of $Na^+/H^+$ antiport activity at the tonoplasts by growth and salt. *Plant Physiology*, **83**, 884–7.

Bourgeais, P., Guerrier, G. & Strullu, D.C. (1987). Adaptation au NaCl de *Lycopersicon esculentum*: étude comparative des cultures de cals ou de parties terminales de tiges. *Canadian Journal of Botany*, **65**, 1989–97.

Carpita, N., Sabularre, D., Montezinos, D. & Pelrer, D.P. (1979). Determination of pore size of cell walls of living plant cells. *Science*, **201**, 1144–7.

Chandler, S.F. & Thorpe, T.A. (1987). Characterization of growth, water relations, and proline accumulation in sodium sulfate tolerant callus of *Brassica napus* L. cv. Westar (Canola). *Plant Physiology*, **84**, 106–11.

Cleland, R.E. (1986). The role of hormones in wall loosening and plant growth. *Australian Journal of Plant Physiology*, **13**, 93–103.

Cocking, E.C. (1986). Somatic hybridization: implications for agriculture. In *Biotechnology in Plant Science*, eds. M. Zaitland, P. Day & A. Hollaender, pp. 101–13. London: Academic Press.

Croughan, T.P., Stavarek, S.J. & Rains, D.W. (1978). Selection of a NaCl tolerant line of cultured alfalfa cells. *Crop Science*, **18**, 959–63.

Cutler, J.M., Rains, D.W. & Loomis, R.S. (1977). The importance of cell size in the water relations of plants. *Physiologia Plantarum*, **40**, 255–60.

Davies, W.J., Metcalfe, J., Lodge, T.A. & daCosta, A.R. (1986). Plant growth substances and the regulation of growth under drought. *Australian Journal of Plant Physiology*, **13**, 105–25.

Dix, P.J. & Pearce, R.S. (1981). Proline accumulation in NaCl-resistant and sensitive cell lines of *Nicotiana sylvestris*. *Zeitschrift für Pflanzenphysiologie*, **102**, 243–8.

Dracup, M., Gibbs, J., Stuiver, C.E.E., Greenway, H. & Flowers, T.J. (1986).

Determination of free space, growth, solute concentration and parameters of water-relations of suspension-cultivated tobacco cells. *Plant, Cell & Environment*, **9**, 693–701.

Duncan, D.R. & Widholm, J.M. (1987). Proline accumulation and its implication in cold tolerance of regenerable maize callus. *Plant Physiology*, **83**, 703–8.

Flowers, T.J. & Yeo, A.R. (1986). Ion relations of plants under drought and salinity. *Australian Journal of Plant Physiology*, **13**, 75–91.

Grumet, R. & Hanson, A.D. (1986). Genetic evidence for an osmoregulatory function of glycinebetaine accumulation in barley. *Australian Journal of Plant Physiology*, **13**, 353–64.

Handa, S., Bressan, R.A., Handa, A.K., Carpita, N.C. & Hasegawa, P.M. (1983). Solutes contributing to osmotic adjustment in cultured plant cells adapted to water stress. *Plant Physiology*, **73**, 834–43.

Handa, S., Handa, A.T., Hasegawa, P.M. & Bressan, R.A. (1986). Proline accumulation and the adaptation of cultured plant cells to stress. *Plant Physiology*, **80**, 938–45.

Hanson, A.D. & Hitz, W.O. (1982). Metabolic responses of mesophytes to plant water deficits. *Annual Review of Plant Physiology*, **33**, 163–203.

Harms, C.T. & Oertli, J.J. (1985). The use of osmotically adapted cell cultures to study salt tolerance *in vitro*. *Journal of Plant Physiology*, **120**, 29–38.

Harrison, B.D. & Mayo, M.A. (1983). The use of protoplasts in plant virus research. In *Use of Tissue Culture and Protoplasts in Plant Pathology*, ed. J.P. Hegelson and B.J. Deverall, pp. 69–129. New York: Academic Press.

Hasegawa, P.M., Bressan, R.A. & Handa, A.K. (1987). Cellular mechanisms of salinity tolerance. *HortScience*, **21**, 1317–24.

Hasegawa, P.M., Bressan, R.A., Handa, S. & Handa, A.K. (1984). Cellular mechanisms of tolerance to water stress. *HortScience*, **19**, 371–7.

Heyser, J.W. & Nabors, M.W. (1981). Growth, water content and solute accumulation of two tobacco cell lines cultured on sodium chloride, dextran and polyethylene glycol. *Plant Physiology*, **68**, 1454–9.

Imbrie-Milligan, C.W. & Hodges, T.K. (1986). Microcallus formation from maize protoplasts prepared from embryogenic callus. *Planta*, **168**, 395–401.

Jefferies, R.L. (1980). The role of organic solutes in osmoregulation in halophytic higher plants. In *Genetic Engineering of Osmoregulation*, ed. D.W. Rains, R.C. Valentine and A. Hollaender, pp. 135–54. New York: Plenum Press.

Jones, H., Leigh, R.A., Tomos, A.D. & Wyn Jones, R.G. (1987). The effect of abscisic acid on cell turgor presures, solute content and growth of wheat roots. *Phytochemistry*, **170**, 257–63.

Koch, G., Wells, R. & Grisebach, H. (1987). Differential induction of enzyme in soybean cell cultures by elictor or osmotic stress. *Planta*, **171**, 519–24.

Kumar, A. & Cocking, E.C. (1987). Protoplast fusion: A novel approach to organelle genetics in higher plants. *American Journal of Botany*, **74**, 1289–303.

La Rosa, P.C., Hasegawa, P.M., Rhodes, D., Clithero, J.M., Watad, A.-E.A. & Bressan, R.A. (1987). Abscisic acid stimulated osmotic adjustment and its involvement in adaptation of tobacco cell to NaCl. *Plant Physiology*, **85**, 174–81.

Leigh, R.A. (1983). Methods, progress and potential for use of isolated vacuoles in studies of solute transport in higher plant cells. *Physiologie Plantarum*, **57**, 390–6.

Leigh, R.A. & Tomos, A.D. (1983). An attempt to use isolated vacuoles to determine the distribution of sodium and potassium in cells of storage roots of red beet (*Beta vulgaris*). *Planta*, **159**, 769–75.

Leigh, R.A. & Wyn Jones, R.G. (1984). A hypothesis relating critical potassium concentration for growth to the distribution and function of this ion in the plant cell. *New Phytologist*, **97**, 1–14.

Lin, W., Odell, J.T. & Schreiner, R.M. (1987). Soybean protoplast culture and direct gene uptake and expression by cultured soybean protoplasts. *Plant Physiology*, **84**, 856–61.

Lone, M.I., Kueh, J.S.H., Wyn Jones, R.G. & Bright, S.W.J. (1987). Influence of proline and glycinebetaine on salt tolerance of cultured barley embryos. *Journal of Experimental Botany*, **38**, 479–90.

Lynch, J., Cramer, G.R. & Läuchli, A. (1987). Salinity reduces membrane-associated calcium in corn root protoplasts. *Plant Physiology*, **83**, 390–4.

Maddock, S.W. & Semple, J.T. (1986). Field assessment of somaclonal variation in wheat. *Journal of Experimental Botany*, **37**, 1114–28.

Moftah, A.E. & Michel, B.E. (1987). The effect of sodium chloride on solute potential and proline accumulation in soybean leaves. *Plant Physiology*, **83**, 238–43.

Paleg, L.G., Stewart, G.R. & Bradbeer, J.W. (1984). Proline and glycine betaine influence protein solvation. *Plant Physiology*, **75**, 974–8.

Pandey, S. & Ganapathy, P.S. (1985). The proline enigma: NaCl-tolerant and NaCl-sensitive callus lines of *Cicer arietinum* L. *Plant Science*, **40**, 13–17.

Pantoja, O. & Willmer, C.M. (1986). Pressure effects on membrane potentials of mesophyll cell protoplasts and epidermal cell protoplasts of *Commelina communis*. *Journal of Experimental Botany*, **37**, 315–20.

Passioura, J.B. (1986). Resistance to drought and salinity: Avenues for improvement. *Australian Journal of Plant Physiology*, **13**, 191–201.

Rains, D.W. (1972). Salt transport by plants in relation to salinity. *Annual Review of Plant Physiology*, **23**, 367–88.

Rains, D.W., Croughton, S.S. & Croughan, T.P. (1986). Isolation and characterization of mutant cell lines and plants: Salt tolerance. In *Cell Culture and Somatic Cell Genetics of Plants*, Vol. 3, ed. I. Vasil, pp. 537–47. New York: Academic Press.

Ramagopal, S. (1987). Salinity stress induced tissue-specific proteins in barley seedlings. *Plant Physiology*, **84**, 324–31.

Reaney, M.J.T. & Gusta, L.V. (1987). Factors influencing the induction of freezing tolerance by abscisic acid in cell suspension cultures of *Bromus inermis* Leyss and *Medicago sativa* L. *Plant Physiology*, **83**, 423–7.

Reuveni, M., Colombo, R., Lerner, H.-R., Pradet, A. & Polkajoff-Mayber, A. (1987). Osmotically induced proton extrusion from carrot cells in suspension culture. *Plant Physiology*, **85**, 383–8.

Rosen, A. & Tal, M. (1981). Salt tolerance in the wild relatives of the cultivated tomato: responses of naked protoplasts isolated from leaves of *Lycopersicon esculentum* and *L. peruvianum* to NaCl and proline. *Zeitschrift für Pflanzenphysiologie*, **102**, 91–4.

Sachs, M.M., Freeling, M. & Okimoto, R. (1980). The anaerobic proteins of maize. *Cell*, **20**, 761–7.

Scowcroft, W.R., Larkin, P.J. & Brettell, R.I.S. (1983). Genetic variation from tissue culture. In *Use of Tissue Culture and Protoplasts in Plant Pathology*, ed. J.P. Helgeson and B.J. Severall, pp. 139–62. New York: Academic Press.

Singh, N.K., Bracken, C.A., Hasegawa, P.M., Handa, A.K., Buckel, S., Hermodson, M.A., Pfankoch, E., Regnier, F.E. & Bressan, R.A. (1987*a*). Characterization of Osmotin. A thaumatin-like protein associated with osmotic adjustment in plant cells. *Plant Physiology*, **85**, 529–36.

Singh, N.K., Handa, A.K., Hasegawa, P.M. & Bressan, R.A. (1985). Proteins associated with adaptation of cultured tobacco cells in NaCl. *Plant Physiology*, **79**, 126–37.

Singh, N.K., La Rosa, P.C., Handa, A.K., Hasegawa, P.M. & Bressan, R.A. (1987*b*). Hormonal regulation of protein synthesis associated salt tolerance in plant cells. *Proceedings of the National Academy of Sciences, USA*, **84**, 739–43.

Spiegel-Roy, M.P. & Ben Hayyim, G. (1985). Selection and breeding for salinity tolerance *in vitro*. *Plant and Soil*, **89**, 243–52.

Stavarek, S.J. & Rains, D.W. (1984*a*). The development of tolerance to mineral stress. *HortScience*, **19**, 377–82.

Stavarek, S.J. & Rains, D.W. (1984*b*) Cell culture techniques: Selection and physiological studies of salt tolerance. In *Salinity Tolerance in Plants: Strategies for Crop Improvement*, ed. R.C. Staples and G.H. Toeniessen, pp. 321–34. New York: Wiley.

Steponkus, P.J. (1980). A unified concept of stress in plants? In *Genetic Engineering of Osmoregulation*, ed. D.W. Rains, R.C. Valentine and A. Hollaender, pp. 235–58. London: Plenum Press.

Stewart, G.R. & Lahrer, F. (1980). Accumulation of amino acids and related compounds in relation to environmental stress. *Biochemistry of Plants*, Vol. 5, ed. B.J. Miflin, pp. 609–35. New York: Academic Press.

Wagner, G.J. (1983). Higher plant vacuoles and tonoplasts. In *Isolation of Membrane and Organelles from Plant Cells*, ed. J.L. Hall and A.L. Moore, pp. 83–118. London: Academic Press.

Watad, A.-E., Rheinhold, L. & Lerner, H.R. (1983). Comparison between a stable NaCl-selected *Nicotiana* cell line and the wild type $K^+$, $Na^+$, and proline pools as a function of salinity. *Plant Physiology*, **73**, 624–9.

Wyn Jones, R.G. (1980). An assessment of quarternary ammonium and related compounds as osmotic effectors in crop plants. In *Genetic Engineering of Osmoregulation*, ed. D.W. Rains, R.C. Valentine and A. Hollaender, pp. 155–70. New York: Plenum Press.

Wyn Jones, R.G. (1984). Phytochemical aspects of osmotic adaptation. In *Recent Advances in Phytochemistry*, Vol. 13, *Phytochemical Adaption to Stress*, ed. B.N. Timmerman, C. Steelink and F.A. Loewus, pp. 55–78. New York: Plenum Press.

Wyn Jones, J.R. & Gorham, J. (1983). Aspects of salt and drought tolerance. In *Genetic Engineering of Plants: An Agricultural Perspect*, ed. T. Kosuge, C.P. Meredith and A. Hollaender, pp. 355–70. New York: Plenum Press.

Yeo, A.R. & Flowers, T.J. (1986). Salinity resistance in rice (*Oryza sativa* L.) and a pyramiding approach to breeding varieties for saline soils. *Australian Journal of Plant Physiology*, **13**, 161–73.

# 11 Breeding methods for drought resistance

## Introduction: improvement of drought resistance in conventional breeding programmes

Critical evaluation of progress in plant breeding over a period of several decades (e.g. Wilcox *et al.*, 1979; Castleberry, Crum & Krull, 1984) has demonstrated a genetic improvement in yield under both favourable and stress conditions. The yield improvement under drought stress occurred before many of the physiological issues of drought resistance were understood and resulted partly from the genetic improvement of yield potential and partly from the improvement of stress resistance. For example, Bidinger *et al.* (1982) found that the yield of millet varieties under drought stress was largely explained by their yield potential and growth duration. Early varieties with a high yield potential were most likely to yield best under stress. Fischer & Maurer (1978) also recognised the effect of potential yield on yield performance of wheat under drought stress and proposed a 'susceptibility index' ($S$) which estimated the relative susceptibility of a variety to drought stress. In analysing their wheat data, they found that $S$ was not totally independent of the potential yield of the variety.

The improvement of yield under stress must therefore combine a reasonably high yield potential (Blum, 1983) with specific plant factors which would buffer yield against a severe reduction under stress. On the other hand, potentially lower yielding genotypes occasionally have been found to perform very well under drought stress conditions (e.g. Blum, 1982; Ceccarelli, 1987), especially under severe drought stress. One is left with the long-standing practical conclusion of Reitz (1974) that 'Varieties fall into three categories: (a) those with uniform superiority over all environments; (b) those relatively better in poor environments; and (c) those relatively better in favoured environments'.

The development of better varieties for dry conditions conventionally involves extensive selection and testing for yield performance over diverse environments using various biometrical approaches (Hanson & Robinson, 1963). These empirical methods have led to the development of drought-resistant cultivars of many crops such as wheat, barley, rice, maize, sorghum and soybean. The intricate statistical and biometrical designs necessary for this approach are required for two reasons. First, the inheritance of yield, which is the major selection criterion in such programmes, is 'complex' as it is determined by a multitude of physiological, biochemical and metabolic plant processes, whose genetics are largely unknown. Even their exact association with plant productivity is unclear. Secondly, the various environments under which genotypes are tested and under which their stability is measured are largely undefined in the biological sense. Breeders are rarely able to identify the main factors that reduce yield in a given environment or their order of importance. Therefore, with the conventional breeding approach the environment is taken as a random occurrence of infinitesimal probabilities, fitting a statistical approach. The level of plant strain in a given environment is measured by the mean yield of the population grown in it, compared with other environments. The biological effects of the different undefined environments are integrated into one statistical parameter of stability in yield performance over changing environments (Finlay & Wilkinson, 1963). Drought resistance is evidently one component of this stability and in cases where environments are variable mainly in the water regime, stable genotypes could indeed be physiologically classified as drought resistant (Blum, 1983; Blum, Mayer & Golan, 1983b; Agarwal & Sinha, 1984).

The main difficulties in the current approach are therefore the use of yield (a genetically 'complex' trait) as a principal selection index and the use of biologically 'invisible' environments for selection towards general stability. A further problem is that the heritability of yield is reduced under conditions where yield is reduced, namely under conditions of stress (e.g. Johnson & Frey, 1967; Rumbaugh, Asay & Johnson, 1984). Selected genotypes that appear to yield relatively well under stress in one cycle of selection may not perform as well in the next cycle, because a large fraction of the apparent variation in the population under stress is environmental rather than genetic. Therefore, selection for drought resistance by the use of yield as a selection index under low-yielding (drought) conditions is inefficient. Furthermore, selection for yield under drought stress must be performed in an environment that corresponds to the drought conditions of the target environment. However, large temporal variability is inherent to the target drought environment and the breeder is in a dilemma as to what kind of selection environment represents the target environment (Rumbaugh et al., 1984).

In order to overcome the low efficiency of such selection programmes, some plant breeders resort to the expensive solution of growing huge populations and performing repeated testing over years and locations to secure results. In such programmes the probability for success becomes a function of investments rather than a function of science.

Although drought-resistant varieties have been developed by the use of empirical breeding methods that employ yield as a selection index, these methods are too costly and require a long period of testing and evaluation. Further improvements by such methods will entail still greater investment. Therefore breeding for drought resistance must depart from the use of yield as the exclusive selection index. A direct reference to some physiological attributes in the selection for drought resistance would allow us to address the underlying factors of stability, to the same extent that selection for, say, disease resistance addresses the specific plant interaction with the pathogen rather than yield. When physiological attributes are addressed, the genetics involved should become simpler, as a function of the lower level of plant organisation. Physiological selection criteria must therefore supplement yield evaluations.

The remainder of this chapter discusses the application to breeding for drought resistance of some of the physiological principles that have been outlined in earlier chapters or elsewhere (Aspinall & Paleg, 1981; International Rice Research Institute, 1982; Blum, 1988).

### Physiological selection criteria

The physiological approach to selection depends on two conditions: (i) an understanding of the physiological parameters of plant water relations which support plant productivity under stress, and (ii) a technical ability for a rapid and simple measurement of the relevant attributes in large breeding populations. Plant breeding is a 'high-tech' industrial endeavour that utilises various scientific disciplines, ranging from botany and chemistry to computer and food sciences. This 'industry' processes raw genetic materials, along engineered production lines, in order to produce at a reasonable cost a competitive product called 'cultivar'. This industry can use only those methods that are economically and logistically compatible with the production line.

Crop plant responses to drought stress in an agricultural ecosystem and the respective physiologies can be classified into two domains. This classification is convenient for understanding the possible practical role of the various drought resistance attributes in economic plants. The first domain is when water stress exists but a positive carbon balance is maintained by the plant. Resistant genotypes achieve a relatively greater net gain of carbon as compared with susceptible ones, with a subsequent effect on yield. The second domain is when stress is severe and a net loss of carbon takes place so that plants are merely surviving stress. Resistant genotypes survive and subsequently recover better upon rehydration, compared with susceptible ones. The physiological and agronomical repercussions of plant responses in these two domains are different and they depend largely upon the nature of drought stress and its profile. Any profile of moderate stress, where yield is not severely reduced, involves the first domain. Severe stresses that terminate growth involve the second domain.

### The domain of net carbon gain

The major physiological attributes that support net carbon gain under conditions of drought stress can be approached by two well-known and simple relationships. The first is that of de Wit (1958),

$$\text{biomass} = mT/E_0 \tag{1}$$

where $T$ is crop transpiration, $E_0$ is potential crop evapotranspiration and $m$ is a plant-specific coefficient. $T$ may be replaced by $E_a$ (actual evapotranspiration) which would then widen the association to include additional effects of canopy architecture and soil surface attributes. Various modifications and derivations have been suggested for this basic relationhip (e.g. Hanks, 1983), but it still serves well for this discussion. The other relationship is that proposed by Passioura (1983),

$$YE = E_a \times WUE \times HI \tag{2}$$

where economic yield ($YE$) equals actual evapotranspiration (crop water use) times water-use efficiency ($WUE$) times harvest index ($HI$).

Therefore drought resistance (in terms of plant production) is positively associated with $T$ or $E_a$, $m$ or $WUE$ and $HI$ under drought stress. A simple simulation using (1) would indicate that a given increase in $T$ or $m$ is most effective towards plant production when water stress is not severe.

### Maintaining transpiration under drought stress

Factors that maintain transpiration under stress will favour continued dry-matter production. However, over the life of a plant optimisation rather than maximisation of water use is required especially during a prolonged period of stress. Optimisation implies some regulation of the use of a limited amount of soil moisture throughout the crop season or the maximisation of soil moisture extraction, when rainfall is expected to follow. For example, optimal use of a given amount of soil moisture is achieved by an optimal stomatal control over leaf gas exchange (Jones,

1981; McCree & Richardson, 1987) or by increased root axial resistances (Passioura, 1983).

*Root growth.* It is necessary to maximise the available water so root growth and development are evidently critical, especially under conditions of high soil resistance to root growth such as in heavy and swelling clay soils or in soils with a hardpan. For some crops such as upland rice, which are characterised by a potentially shallow root system, appreciable gains in drought resistance have been achieved by the genetic improvement of root extension (e.g. IRRI, 1982). Even for crops such as sorghum, with a relatively highly developed root system, a clear advantage for extensive rooting was apparent where soil resistance to root penetration was high (Jordan *et al.*, 1983). The ability of some modern soybean cultivars better to withstand drought stress by dehydration avoidance was ascribed to the genetic improvement of root growth (Boyer, Johnson & Saupe, 1980). It appears that root growth is slowly being optimised within the framework of the conventional breeding programme, under the effect of repeated selection and testing for yield stability over different environments. Indeed, it has been concluded for sorghum that large genetic variation for root growth could be revealed only within newly developed exotic breeding stocks, while all current varieties tested were nearly the same in this respect (Jordan & Miller, 1980).

Root growth is also affected by genetic factors that are not associated directly with the control of roots. For example, genes that control plant maturity of sorghum affect root growth rate and root length density (Blum, Arkin & Jordan, 1977; Blum & Arkin, 1984). Furthermore, the adaptive responses of the shoot to drought stress may also affect root growth (Sharp & Davies, 1979; McGowan *et al.*, 1984; Morgan & Condon, 1986). On the other hand, whatever the internal plant control over root growth may be, the soil environment has a striking effect on root development (IRRI, 1982; Blum & Ritchie, 1984). Although potential root growth in the absence of root resistance may be immense (as seen in root culture experiments in aeroponics) genotypes interact with soil characteristics in their rooting behaviour. In view also of the inherent difficulties in measuring roots and the effect of the soil environment on roots, the direct treatment of roots in selection for drought resistance is impractical but, as has been shown for rice (IRRI, 1982) and maize (Fischer, Johnson & Edmeades, 1983), the selection for a better leaf water status under stress may result in improved root development.

Some modern varieties of cereals such as wheat and rice achieve high yield potential through genetic factors for plant dwarfness that improve harvest index. A popular notion among plant breeders is that high-yielding dwarf varieties are more susceptible to drought stress. Their suspected susceptibility to drought stress is suggested to result from a possible reduction in root growth in association with dwarfness. However, studies performed with nearly isogenic lines of cereals differing for height genes have not substantiated the assumption that dwarf genotypes generally lack in root growth (e.g. Pepe & Welsh, 1979; Owonubi & Kanemasu, 1982; Fischer & Wood, 1979).

*Osmoregulation.* The role of osmoregulation in plant resistance to environmental stresses that involve water stress components is discussed in detail in Chapters 5, 6 and 12 (see also Blum, 1988). For the purposes of this chapter it is worth noting that indirect evidence on the role of osmo-regulation in drought resistance and crop plant production comes from several sources. Morgan (1983) was able to select wheat plants with high or low osmoregulation under stress. He demonstrated that high-osmoregulat-ing genotypes yielded better under dryland conditions than did low-osmoregulating genotypes and that the genetics of osmoregulation was likely to be relatively simple. Other work in wheat (Blum *et al.*, 1983b) suggested that a drought-resistant wheat variety, in terms of its long-term yield stability under drought stress, was characterised by a greater capacity for osmoregulation compared with less resistant varieties. Landraces of sorghum and millet from dry regions in India and Africa were found to be more drought resistant (in terms of plant growth and delayed leaf sene-scence) than landraces from humid regions (Blum & Sullivan, 1986). The races from dry regions all excelled in osmoregulation. Osmoregulation and turgor maintenance permit continued root growth and thorough soil moisture extraction (see detailed discussion in Sharp & Davies, 1979; Chapter 5). McCree, Kallsen & Richardson (1984) showed that osmo-regulation enabled the maintenance of a positive carbon balance during stress in sorghum and that carbon stored during stress was immediately available for growth upon rehydration and recovery. While the role of osmoregulation largely involves turgor maintenance, other effects of osmo-regulation have been proposed; these include the protection of membranes against functional or structural perturbations by stress (Chapter 6; Blum, 1988).

On some occasions osmoregulation may be linked with reduced growth across variable genetic materials, such as in exotic cotton germplasm (Quisenberry, Cartwright & McMichael, 1984). This has been confirmed by work with sorghum (Blum & Sullivan, 1986) and wheat (A. Blum, unpublished) seedlings, where drought resistance and osmoregulation under stress were greater in varieties of potentially smaller plants.

Attempts to select osmoregulating cell lines *in vitro* as means for improving drought (and salinity) resistance have been reviewed in Chapters 10 & 12, but it is worth noting here that, as with other *in vitro* selection programmes for whole-plant responses to environmental stresses, difficulties remain in recovering durable resistance in plantlets and plants. There is also the serious question of the extent of association in resistance between cells and the whole plant. While osmoregulation is indeed a cellular phenomenon, it is quite possible that osmoregulation in the intact plant may be achieved through modified assimilate partitioning among organs (Blum, Mayer & Golan, 1988*b*), tissues and cells. The current understanding of the dynamics of plant response to water deficit favours the assumption that some growth retardation (by genetical or environmental factors), as suggested above, would release assimilates for osmoregulation. Such associations can hardly be considered in selection *in vitro*.

The direct measurement of osmoregulation in large breeding populations is practically impossible. Although Morgan (1983) was able to measure a large number of genotypes by thermocouple psychrometry, this slow method would be unacceptable in routine breeding programmes. Selection for the maintenance of turgor and/or transpiration by osmoregulation or by any other plant factor can proceed and is being practised through several rapid indirect methods.

*Canopy temperature.* Genotypic variations in transpiration under stress can be monitored by the remote sensing of canopy temperature with an infra-red thermometer because high transpiration rates are generally reflected in lower canopy temperatures. The method has been applied successfully to the selection of wheat (Blum, Mayer & Golan, 1982) and maize (Fischer *et al*, 1983) under drought stress. Selected progeny of maize plants with cooler leaves under stress were found to yield better under stress because of their greater root growth and better soil moisture extraction. Additional evidence for the value of the technique has been obtained for soybeans (Harris, Schapaugh & Kanemasu, 1984), sorghum (Chaudhuri *et al.*, 1986) and cotton (Hatfield, Quisenberry & Dilbeck, 1987). Selection can be based on the relative differences obtained in canopy temperatures between genotypes at the same time, provided that plants are well stressed (Blum *et al.*, 1982). It is sufficient in selection work to classify all materials into high, medium and low temperature classes. Generally, genotypes with lower temperatures are preferable, though genotypes with higher temperatures (indicating lower transpiration) may be advantageous when water must be preserved for later and for sensitive developmental stages under very dry conditions (Hatfield *et al.*, 1987). This underlines again that control rather than maximisation of transpiration is the important issue.

*Leaf firing.* Rapid senescence of leaves ('leaf firing') is a well-known symptom of water stress and indicates leaf tissue death resulting from the high leaf temperatures reached when transpiration ceases. In most crop plants leaf killing temperatures are within a range of 45 to 55 °C. Leaf firing may be taken as a simple visual criterion for drought injury and has been used extensively in the selection for drought resistance in maize (Castleberry, 1983; Fischer *et al.*, 1983) and sorghum (Rosenow, Quisenberry & Wendt, 1983). However, since observations at later growth stages may also involve an effect on natural senescence, such scores should be normalised for plant phenology.

*Leaf rolling.* Leaf rolling is a well recognised symptom of water stress in the cereals (Hsiao *et al.*, 1984), while specific leaf movements as a reflection of turgor loss also occur in the legumes. It has often been observed that some genotypes roll their leaves more readily than others under drought stress, though genotypic differences in leaf morphology can have a large effect on the degree of leaf rolling (Jones, 1979). Reduced leaf rolling in some rice genotypes was found to result from better osmoregulation (Hsiao *et al.*, 1984), while reduced leaf rolling in some rice (O'Toole & Moya, 1978; Jones, 1979), wheat (Jones, 1979) and sorghum (A. Blum, unpublished) genotypes was related to the maintenance of a higher leaf water potential. Leaf rolling has often been suggested to be a positive attribute under stress as it may reduce the energy load on the leaf, but whatever the actual ecological significance of leaf rolling, its relative delay can be taken as an indirect estimate of turgor maintenance, either through osmoregulation or through the maintenance of a higher leaf water potential in the field. Leaf rolling is therefore being visually scored and used in the selection for drought resistance in rice (IRRI, 1982), sorghum (Rosenow *et al.*, 1983) and maize (Sobrado, 1987).

### The m *coefficient and* WUE.

Data compiled by Hanks (1983) show that, contrary to earlier notions, the *m* coefficient in Equation 1 may be affected by the genotype. A relatively greater net gain of carbon for the same rate of transpiration under stress may be reflected in the *m* coefficient (Equation 1), in the ratio between assimilation and transpiration (assimilation ratio) or in the agronomic index, *WUE* (Equation 2).

The question whether assimilation ratio or crop *WUE* can be genetically modified within any species is at present under extensive debate. Genetic variations seen at the single leaf or single plant level could not be verified at the field level (e.g. Jarvis & McNaughton, 1986). The following points should serve to indicate that there may be a real potential for genetically modifying crop water-use efficiency.

*Leaf wax.* Since the early investigations of plant xerophytism, heavy epicuticular wax load on leaves has been believed to provide some drought resistance by decreasing cuticular water loss and therefore improving stomatal control over water loss (Blum, 1988). Genetic variation for epicuticular wax load in crop plants has been noted (e.g. Ebercon, Blum & Jordan, 1977; O'Toole, Cruz & Sieber, 1979; Jordan, Douglas & Shouse, 1983), but leaf wax is also affected by environmental factors (e.g. Jordan *et al.*, 1983) such as water stress, high temperature and high radiation. Increased cuticular resistance due to the presence of epicuticular wax was indeed reflected in an increase in the assimilation ratio of sorghum (Chatterton *et al.*, 1975). Epicuticular wax also affects the spectral properties of leaves (Blum, 1975) to the extent that net radiation and leaf temperature may be reduced. The effect of glaucousness in wheat was found to be associated largely with reduced leaf temperatures and delayed heat-induced leaf senescence (Johnson, Richards & Turner, 1983). Epicuticular wax can be determined by both gravimetric and rapid colorimetric methods. However, in many plant species it can easily be identified visually by the glaucous appearance of the plants or parts thereof (e.g. Johnson *et al.*, 1983).

*Awns.* Assimilation ratio may differ among plant organs. Data for wheat and barley showed (Blum, 1985) that net photosynthesis per unit area in awns was lower than that in glumes or in the flag leaf. However, transpiration in awns was disproportionately lower than in glumes or in the flag leaf. Thus, the assimilation ratio in awns was greater by several orders of magnitude than that in the glumes or flag leaves (Blum, 1986). Long awns are therefore a drought-adaptive feature in the cereals, and one that may increase *WUE* after anthesis. It is therefore perhaps not a coincidence that awned genotypes of sorghum had a selective advantage in mass selection under dry conditions but not under irrigated conditions (Keim, Miller & Rosenow, 1983). Cereal genotypes vary greatly in awn size. The attribute is relatively simply inherited and selection for the number or size of awns is a simple task that may be done visually.

*Carbon assimilation.* Cellular and subcellular attributes that are directly associated with the maintenance of carbon gain under stress are not commonly used as selection indices, mainly because of methodological difficulties even though, for example, the performance of the photosynthetic apparatus under conditions of cellular stress is of primary importance. The measurement of gas exchange in various genotypes subjected to the same level of water deficit is slow and impractical in selection work, but use of chlorophyll fluorescence as a fast probe of photosystem II responses (Havaux & Lannoye, 1985) may hold a potential in selection for drought resistance. The recent finding that the leaf isotopic carbon ratio ($^{13}C/^{12}C$) can be predictive of the genotype's *WUE* is potentially extremely important (Farquhar & Richards, 1984; Hubick, Farquhar & Shorter, 1986) and is discussed in detail in Chapter 4. A particular advantage in the proposed method is that plants do not have to be water stressed when analysed. On the other hand, the analysis requires expensive instrumentation that is not readily available to most breeders, a fact that might limit its wide use in practical selection work. The potential of the method, however, has been demonstrated by the fact that genotypic differences in $^{13}C/^{12}C$ have been correlated with *WUE* in both wheat and groundnut.

### Harvest index (HI) under stress.

In the grain crops, harvest index (the ratio of grain weight to total above-ground biomass) is affected by the size of the reproductive sink and by the rate and duration of assimilate partitioning to this sink. Among the many different physiological processes affected by water stress, translocation has been considered to be relatively tolerant (Hsiao, 1973). In several of the cereals grain growth is supported partly by transient photosynthesis (by the flag leaf and the inflorescence) and partly by translocation of stored stem reserves. Different reports have assigned very different values to the contribution of stem reserves to grain filling. It is evident now (Blum, 1988) that when transient photosynthesis is limited by water stress during grain filling, stem reserve mobilisation assumes a greater relative and absolute role in grain filling. Large stem reserves and a greater capacity for its mobilisation would therefore support *HI* under drought stress. The yields of various cereals under dry conditions have been found to be positively associated with the rate of stem weight loss, as an expression of reserve export (e.g. Rawson, Bagga & Bremner, 1977; Agarwal & Sinha, 1984). While the 'non-senescent' types of sorghum appear to be relatively less water-stressed when drought occurs after anthesis, it was found that the 'senescent' types were translocating more assimilates from the shoot to the grain (Harden & Krieg, 1983). The 'non-senescent' trait, which ascribes an appearance of well-being to plants under a late-season drought stress, may very well be linked with the conservation of carbohydrates in the shoot when the grain develops, which is undesirable in terms of maximising *HI*.

Selection for the capacity of grain filling by stem reserve mobilisation under a post-anthesis drought stress is difficult. It is nearly impossible to apply the proper selection pressure by affecting the same level of water stress during grain filling over a phenologically variable germplasm. A simple indirect method has been developed for the small grains in order to reveal the capacity for grain filling in the absence of transient photosynthesis (Blum, Mayer & Golan, 1983a, b). With this method plants are grown in the field under well-watered conditions. The photosynthetic source is destroyed after anthesis by spraying the plants with a 4% solution of magnesium (or sodium) chlorate. The chemical desiccates and bleaches the plants without killing them. The spray is applied to all genotypes at the same developmental stage (14 days after anthesis), when grain growth is into its exponential phase and the demand for assimilates is at its height. Genotypes are then evaluated at maturity by comparing kernel weight between sprayed and non-sprayed plots, and by calculating kernel weight loss caused by the treatment. Diverse wheat materials, for example, differed within a range of 5–50% loss by this treatment, and the response in kernel weight loss to the treatment was correlated with the response to a late-season drought, over genotypes. It was also found that this screening method was efficient for revealing tolerance to other stress factors that inhibit photosynthesis during grain filling, such as certain leaf diseases (Zilberstein, Blum & Eyal, 1985).

### The domain of net carbon loss

Surprisingly, very little physiological work has been done to understand the nature and processes of plant recovery from extreme drought stress, especially in relation to plant production (Chapter 7). In order for the plant to recover properly from severe water stress, its various meristems must survive. The association between severe plant stress and the factors that affect meristem survival and function upon rehydration are unclear though osmoregulation may have a possible protective role and as a potential source of carbon for recovery. Active plant apices generally excel in osmoregulation and do not lose much water upon plant dehydration (Barlow, Munns & Brady, 1980).

The ability to recover is a function of plant age. This can be seen in the general loss of cellular drought tolerances during plant ontogeny with the seed and the initial stages of germination being the most tolerant, followed by juvenile plants and meristems. Thus it has often been observed that late-maturing genotypes or plants of indeterminate growth habit are more likely to recover better than earlier or determinate plants. Early-maturing genotypes have an advantage under conditions of limited water supply and late-season water stress, as their water requirement is smaller at most growth stages (Blum & Arkin, 1984). On the other hand, late-maturing genotypes have an advantage under conditions of severe mid-season water stress, followed by a late-season rehydration period.

In the reality of an agricultural ecosystem, successful plant recovery after stress is linked with several agronomic variables including the length of the effective growing season. If recovery is delayed it may then take place in an unsuitable environment, in terms of photoperiod or temperature. Thus, in spite of a good recovery, yield may still be very low. Often plants are planted early enough in order to escape late-season disease, pest or climatic problems. Delayed recovery may also subject plants to such problems. Genotypes resistant to such diseases or pests are mandatory for achieving a successful yield after recovery.

Plant regrowth upon recovery undoubtedly involves endogenous hormones. Abscisic acid (ABA) is known to accumulate in leaves of water stressed plants, subsequently causing stomatal closure. The effect of ABA in this respect can be counterproductive in terms of achieving carbon gain during water stress. This is consistent also with the finding that ABA accumulation causes floret sterility in wheat (Morgan, 1980). The association established between a greater capacity for ABA accumulation under stress and drought resistance (e.g. Ilahi & Dörffling, 1982; Ackerson, 1983; Innes, Blackwell & Quarrie, 1984) is therefore difficult to understand, unless the capacity for recovery is considered (see also Chapter 1). If ABA is accumulated and stomata close at a higher leaf water potential in 'high ABA' genotypes, then plant tissues would be less likely to be subjected to extreme desiccation than in 'low ABA' genotypes. Such tissues would therefore be better conserved and more able to recover upon rehydration. Apparently, much more research is needed on plant recovery from stress, and its possible association with hormones.

It has been shown in wheat that recovery by new tillers can lead to very reasonable yields (Hochman, 1982). Similarly there are several published examples of the better recovery of indeterminate genotypes in the legumes. Tests for plant recovery are often employed in the field or in the growth chamber. Differences among genotypes can be outstanding. Visual observations of the amount of new growth of main culms or tillers are sufficient for separating extreme genotypes in this respect, though ranking of genotypes for recovery must be normalised for phenology. Seedling survival after stress is a very relevant issue in range grass breeding, and it is measured directly in the field or under simulated stress conditions (Asay & Johnson, 1983). Seedling survival, however, may not only be related to tissue desiccation problems but also to the rooting behaviour of the seedling. The use of germination rate in osmoticum to predict seedling drought resistance has often been tried in the past with inconsistent results, and it is now clear that seed germination in osmoticum is not even predictive of seed germination in a drying soil (Blum, 1988).

### Conclusions for the breeder

An efficient breeding programme for drought resistance is difficult to achieve by using the sole criterion of yield performance under stress, so physiological selection criteria must be used in selection. Since yield potential has a net effect on yield performance under drought stress, the ideotype must be drought resistant and of a reasonably high potential yield. Scattered circumstantial evidence indicates that to some extent there may exist a trade-off between yield potential and drought resistance. For environments of unpredictable and/or mild drought conditions, potentially high-yielding genotypes that carry some elementary properties of drought adaptation (proper phenology, deeper roots, etc.) may be desirable. For the more consistently harsh environments there may be a need to trade-off yield potential for maximised drought resistance.

The question of drought resistance and potential yield has an important bearing on the search for drought-resistant genetic resources. Undoubtedly, some wild relatives of cultivated crop plants are drought resistant (Quisenberry *et al.*, 1981; Shimshi, Mayoral & Atsmon, 1982; Sobrado & Turner, 1983), but drought resistance of wild species has evolved under natural selection for plant survival. This resistance may not be useful for crop plants where production rather than survival of the species is concerned. This is indirectly supported by the work of Vaadia (1987), who did not find any advantage for components of plant production and water-use efficiency in a drought-resistant wild *Solanum* species. Work in wheat suggests that when the proper tests are used, ample variation for drought resistance may still be found within the routine agronomic wheat breeding germplasm (Blum, Gollan & Mayer, 1981). Even some of the most modern varieties of crop plants have been found to show good levels of drought resistance (e.g. Boyer *et al.*, 1980; Keim & Kronstad, 1981). The most immediate usefulness of exotic genetic resources for drought resistance appears to be in landraces of the various crop plants, as seen for maize (Fischer *et al.*, 1983), sorghum (Blum & Sullivan, 1986) and wheat (Blum *et al.*, 1988a).

In terms of general breeding strategy, first stage selection must still address yield potential and general agronomic traits, before selection for drought resistance is performed. This is consistent with the fact that most practical selection methods for drought resistance do not fit single plant selection and are therefore appropriate at more advanced generations (say $F_4$ and so on). Within the general framework of a breeding programme, suitable materials are advanced to, say, the $F_4$ generation by criteria of yield components, disease reaction, phenology, grain quality, etc. at which stage tests for drought response are initiated (Blum, 1983). A schematic example of the general outline of the wheat breeding programme at our Institute is given in Fig. 1. Materials are advanced to the $F_4$ stage on the basis of selection for general agronomic attributes, when selection for drought resistance by infra-red thermometry, chemical desiccation and yield under stress and non-stress conditions is affected. A summer nursery is used also for the selection for heat tolerance (Shpiler & Blum, 1986).

The insufficiency of a single physiological selection criterion for drought resistance has been indicated repeatedly. The most urgent problem in breeding for drought resistance is the construction of a multiple selection

Fig. 1. A general, simplified and self-explanatory layout of the wheat breeding programme at the Volcani Center, Israel. The selection for drought resistance is affected at about the $F_4$ and later generations, by using infra-red thermometry, chemical desiccation and yield under stress. A summer (off-season) nursery is incorporated as means for selection for heat tolerance.

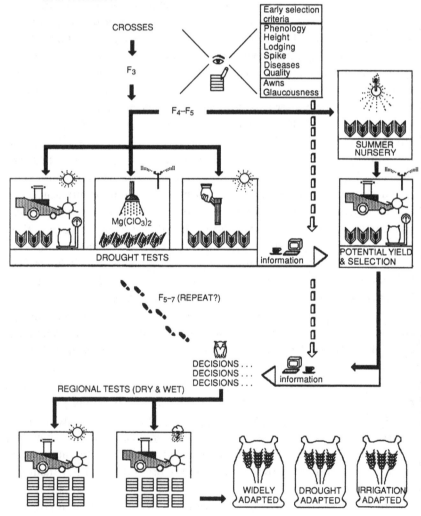

index that takes account of the target environment, the physiological issues involved, the available methodology and the economic constraints of the programme. What can be done today is the result of existing constraints rather than a result of free choice. Recent developments in the direct probing of plant production processes and further research into the physiology and genetics of drought resistance may open the way for upgrading the selection system in the future.

### References

Ackerson, R.C. (1983). Comparative physiology and water relations of two corn hybrids during water stress. *Crop Science*, **23**, 278–83.

Agarwal, P.K. & Sinha, S.K. (1984). Effect of water stress on grain growth and assimilate partitioning in two cultivars of wheat contrasting in their yield stability in a drought-environment. *Annals of Botany*, **53**, 329–40.

Asay, K.H. & Johnson, D.A. (1983). Breeding for drought resistance in range grasses. *Iowa State Journal of Research*, **57**, 441–55.

Aspinall, D. & Paleg, L.G. (ed.) (1981). *The Physiology and Biochemistry of Drought Resistance in Plants*. New York: Academic Press.

Barlow, E.W.R., Munns, R.E. & Brady, C.J. (1980). Drought responses of apical meristems. In *Adaptation of Plants to Water and High Temperature Stress*, ed. N.C. Turner and P.J. Kramer, pp. 191–206. New York: Wiley Interscience.

Bidinger, F.R., Mahalakshmi, V., Talukdar, B.S. & Alagarswamy, G. (1982). Improvement of drought resistance in millet. In *Drought Resistance in Crop Plants with Emphasis on Rice*, pp.357–76. Los Baños, Philippines: International Rice Research Institute.

Blum, A. (1975). Effect of the Bm gene on epicuticular wax deposition and the spectral characteristics of sorghum leaves. *SABRAO Journal*, **7**, 45–9.

Blum, A. (1982). Evidence for genetic variability in drought resistance and its implication in plant breeding. In *Drought Resistance in Crops With Emphasis on Rice*, pp. 52–70. Los Baños, Philippines: International Rice Research Institute.

Blum, A. (1983). Breeding programs for improving crop resistance to water stress. In *Crop Reactions to Water and Temperature Stresses in Humid, Temperate Climates*, ed. C.D. Raper Jr and P.J. Kramer, pp. 263–76. Boulder, Colorado: Westview Press.

Blum, A. (1985). Photosynthesis and transpiration in leaves and ears of wheat and barley varieties. *Journal of Experimental Botany*, **36**, 432–40.

Blum, A. (1986). The effect of heat stress on leaf and ear photosynthesis. *Journal of Experimental Botany*, **37**, 111–18.

Blum, A. (1988). *Plant Breeding for Stress Environments*. Boca Raton: CRC Press.

Blum, A. & Arkin, G.F. (1984). Sorghum root growth and water-use as affected by water supply and growth duration. *Field Crops Research*, **9**, 131–42.

Blum, A., Arkin, G.F. & Jordan, W.R. (1977). Sorghum root morphogenesis and growth. I. Effect of maturity genes. *Crop Science*, **17**, 149–53.

Blum, A., Golan, G., Mayer, J. (1981). The manifestation of dehydration avoidance in wheat breeding germplasm. *Crop Science*, **21**, 495–9.

Blum, A., Golan, G., Mayer, J., Sinmena, B., Shpiler, L. & Burra, J. (1988a). The drought resistance of landraces of wheat from the Northern Negev desert in Israel. *Euphytica*, **28** (in press).

Blum, A., Mayer, J. & Golan, G. (1982). Infrared thermal sensing of plant canopies as a screening technique for dehydration avoidance in wheat. *Field Crops Research*, **57**, 137–46.

Blum, A., Mayer, J. & Golan, G. (1983*a*). Chemical desiccation of wheat plants as a simulator of post-anthesis stress. II. Relations to drought stress. *Field Crops Research*, **6**, 149–55.

Blum, A., Mayer, J. & Golan, G. (1983*b*). Associations between plant production and some physiological components of drought resistance. *Plant, Cell and Environment*, **6**, 219–25.

Blum, A., Mayer, J. & Golan, G. (1988*b*). The effect of grain number per ear (sink size) on source activity and its water-relations in wheat. *Journal of Experimental Botany*, **39**, 106–114.

Blum, A. & Ritchie, J.T. (1984). Effect of soil surface water content on sorghum root distribution in the soil. *Field Crops Research*, **8**, 169–76.

Blum, A. & Sullivan, C.Y. (1986). The comparative drought resistance of landraces of sorghum and millet from dry and humid regions. *Annals of Botany*, **57**, 835–46.

Boyer, J.S., Johnson, R.R. & Saupe, S.G. (1980). Afternoon water deficits and grain yields in old and new soybean cultivars. *Agronomy Journal*, **72**, 981–6.

Castleberry, R.M. (1983). Breeding programs for stress tolerance in corn. In *Crop Reactions to Water and Temperature Stresses in Humid, Temperature Climates*, ed. C.D. Raper Jr and P.J. Kramer, pp. 277–88. Boulder, Colorado: Westview Press.

Castleberry, R.M., Crum, C.W. & Krull, C.F. (1984). Genetic yield improvement of U.S. maize cultivars under varying fertility and climatic environments. *Crop Science*, **24**, 33–7.

Ceccarelli, S. (1987). Yield potential and drought tolerance of segregating populations of barley in contrasting environments. *Euphytica*, **36**, 265–73.

Chatterton, N.J., Hanna, W.W., Powell, J.B. & Lee, D.R. (1975). Photosynthesis and transpiration of the bloom and bloomless sorghum. *Canadian Journal of Plant Science*, **77**, 641–3.

Chaudhuri, U.N., Deaton, M.L., Kanemasu, E.T., Wall, G.W., Marcarian, V. & Dobrenz, A.K. (1986). A procedure to select drought-tolerant sorghum and millet genotypes using canopy temperature and vapor pressure deficit. *Agronomy Journal*, **78**, 490–4.

Ebercon, A., Blum, A. & Jordan, W.R. (1977). A rapid colorimetric method for epicuticular wax content of sorghum leaves. *Crop Science*, **17**, 179–80.

Farquhar, G.D. & Richards, R.A. (1984). Isotopic composition of plant carbon correlates with water-use efficiency of wheat genotypes. *Australian Journal of Plant Physiology*, **11**, 539–52.

Finlay, K.C. & Wilkinson, G.N. (1963). The analysis of adaptation in plant breeding programme. *Australian Journal of Agricultural Research*, **14**, 742–61.

Fischer, K.S., Johnson, E.C. & Edmeades, G.O. (1983). *Breeding and Selection for Drought Resistance in Tropical Maize*. Mexico: International Maize and Wheat Improvement Center.

Fischer, R.A. & Maurer, R. (1978). Drought Resistance in spring wheat cultivars. I. Grain yield responses. *Australian Journal of Agricultural Research*, **29**, 897–905.

Fischer, R.A. & Wood, J.T. (1979). Drought resistance in spring wheat cultivars. III. Yield associations with morpho-physiological traits. *Australian Journal of Agricultural Research*, **30**, 1001–11.

Hanks, R.J. (1983). Yield and water-use relationships: an overview. In *Limitations to Efficient Water Use in Crop Production*, ed. H.M. Taylor, W.R. Jordan and T.R. Sinclair, pp. 393–412. Madison, Wisconsin: American Society of Agronomy.

Hanson, W.D. & Robinson, H.F. (1963). *Statistical Genetics and Plant Breeding*. Washington, DC: National Academy of Sciences, National Research Council.

Harden, M.L. & Krieg, D.R. (1983). Contribution of preanthesis assimilate to grain fill in sorghum. *Cereal Foods World*, **28**, 562–3.

Harris, D.S., Schapaugh, W.T. & Kanemasu, E.T. (1984). Genetic diversity in soybeans for leaf canopy temperature and the association of leaf canopy temperature and yield. *Crop Science*, **24**, 839–42.

Hatfield, J.L., Quisenberry, J.E. & Dilbeck, R.E. (1987). Use of canopy temperature to identify water conservation in cotton germplasm. *Crop Science*, **26**, 269–73.

Havaux, M. & Lannoye, R. (1985). Drought resistance of hard wheat cultivars measured by a rapid chlorophyll fluorescence test. *Journal of Agricultural Science, Cambridge*, **104**, 501–4.

Hochman, Z. (1982). Effect of water stress with phasic development on yield of wheat grown in semi-arid environment. *Field Crops Research*, **5**, 55–68.

Hsiao, T.C. (1973). Plant responses to water stress. *Annual Revivew of Plant Physiology*, **24**, 519–70.

Hsiao, T.C., O'Toole, J.C., Yambao, E.B. & Turner, N.C. (1984). Influence of osmotic adjustment on leaf rolling and tissue death in rice (*Oryza sativa* L.). *Plant Physiology*, **75**, 338–41.

Hubick, K.T., Farquhar, G.D. & Shorter, R. (1986). Correlation between water-use efficiency and carbon isotope discrimination in diverse peanut (*Arachis*) germplasm. *Australian Journal of Plant Physiology*, **13**, 803–16.

Ilahi, I. & Dörffling, K. (1982). Changes in abscisic acid and proline levels in maize varieties of different drought resistance. *Physiologia Plantarum*, **55**, 129–35.

Innes, P., Blackwell, R.D. & Quarrie, S.A. (1984). Some effects of genetic variation in drought-induced abscisic acid accumulation on the yield and water-use of spring wheat. *Journal of Agricultural Research, Cambridge*, **102**, 341–51.

International Rice Research Institute (1982). *Drought Resistance in Crop Plants with Emphasis on Rice*. Los Baños, Philippines: International Rice Research Institute.

Jarvis, P.S. & McNaughton, K.G. (1986). Stomatal control of transpiration: scaling up from leaf to region. *Advances in Ecological Research*, **15**, 1–49.

Johnson, D.A., Richards, R.A. & Turner, N.C. (1983). Yield, water relations, gas exchange and surface reflectance of near isogenic wheat lines differing in glaucousness. *Crop Science*, **23**, 318–21.

Johnson, G.R. & Frey, K.J. (1967). Heritabilities of quantitative attributes of oat (*Avena* sp.) at varying levels of environmental stress. *Crop Science*, **7**, 43–6.

Jones, H.G. (1979). Visual estimation of plant water status in cereals. *Journal of Agricultural Science, Cambridge*, **92**, 83–9.

Jones, H.G. (1981). The use of stochastic modelling to study the influence of stomatal behaviour on yield–climate relationships. In *Quantitative Aspects of Plant Physiology*, ed. D.A. Charles-Edwards and D.A. Rose, pp. 231–40. London: Academic Press.

Jordan, W.R., Douglas, P.R. Jr & Shouse, P.J. (1983). Strategies for crop improvement for drought-prone regions. *Agricultural Water Management*, **7**, 281–99.

Jordan, W.R. & Miller, F.R. (1980). Genetic variability in sorghum root systems: implication for drought tolerance. In *Adaptation of Plants to Water and High Temperature Stress*, ed. N.C. Turner and P.J. Kramer, pp. 383–400. New York: Wiley Interscience.

Jordan, W.R., Monk, R.L. Miller, F.R., Rosenow, D.T., Clark, R.E. & Shouse, P.J. (1983). Environmental physiology of sorghum. I. Environmental and genetic control of epicuticular wax load. *Crop Science*, **23**, 552–6.

Keim, D.L. & Kronstad, W.E. (1981). Drought response of winter wheat cultivars under field stress conditions. *Crop Science*, **21**, 11–15.

Keim, K.R., Miller, F.R. & Rosenow, D.T. (1983). Natural selection under diverse environments for monogenic traits in a sorghum composite. *Agronomy Abstracts*, pp. 69.

McCree, K.J., Kallsen, C.E. & Richardson, S.G. (1984). Carbon balance of sorghum plants during osmotic adjustment to water stress. *Plant Physiology*, **76**, 898–902.

McCree, K.J. & Richardson, S.G. (1987). Stomatal closure vs. osmotic adjustment: a comparison of stress responses. *Crop Science*, **27**, 539–43.

McGowan, M., Blanch, P., Gregory, P.J. & Haycock, D. (1984). Water relations of winter wheat. 5. The root system and osmotic adjustment in relation to crop evaporation. *Journal of Agricultural Science, Cambridge*, **102**, 415–25.

Morgan, J.M. (1980). Possible role of abscisic acid in reducing seed set in water stressed wheat plants. *Nature*, **285**, 655–7.

Morgan, J.M. (1983). Osmoregulation as a selection criterion for drought tolerance in wheat. *Australian Journal of Agricultural Research*, **34**, 607–14.

Morgan, J.M. (1984). Osmoregulation and water stress in higher plants. *Annual Review of Plant Physiology*, **35**, 299–319.

Morgan, M.J. & Condon, A.G. (1986). Water-use, grain yield and osmoregulation in wheat. *Australian Journal of Plant Physiology*, **13**, 523–32.

O'Toole, J.C., Cruz, R.T. & Sieber, J.N. (1979). Epicuticular wax and cuticular resistance in rice. *Physiologia Plantarum*, **47**, 239–43.

O'Toole, J.C. & Moya, T.B. (1978). Genotypic variation in maintenance of leaf water potential in rice. *Crop Science*, **18**, 873–6.

Owonubi, J.J. & Kanemasu, E.T. (1982). Water-use efficiency of three height isolines of sorghum. *Canadian Journal of Plant Science*, **62**, 35–41.

Passioura, J.B. (1983). Roots and drought resistance. *Agricultural Water Management*, **7**, 265–80.

Pepe, J.F. & Welsh, J.R. (1979). Soil water depletion patterns under dryland field conditions of closely related height lines of wheat. *Crop Science*, **19**, 677–81.

Quisenberry, J.E., Cartwright, G.B. & McMichael, B.L. 1984. Genetic relationship between turgor maintenance and growth in cotton germplasm. *Crop Science*, **24**, 479–82.

Quisenberry, J.E., Jordan, W.R., Roark, B.A. & Fryrear, D.W. (1981). Exotic cottons as genetic sources for drought resistance. *Crop Science*, **21**, 889–95.

Rawson, H.M., Bagga, A.K. & Bremner, P.M. (1977). Aspects of adaptation by wheat and barley to soil moisture deficits. *Australian Journal of Plant Physiology*, **4**, 389–401.

Reitz, L.P. (1974). Breeding for more efficient water-use – is it real or a mirage? *Agricultural Water Management*, **14**, 3–28.

Rosenow, D.T., Quisenberry, J.E. & Wendt, C.W. (1983). Drought tolerant sorghum and cotton germplasm. *Agricultural Water Management*, **7**, 207–22.

Rumbaugh, M.D., Asay, K.H. & Johnson, D.A. (1984). Influence of drought stress on genetic variances of alfalfa and wheatgrass seedlings. *Crop Science*, **24**, 297–302.

Sharp, R.E. & Davies, W.J. (1979). Solute regulation and growth by root and shoots of water stressed maize plants. *Planta*, **147**, 43–9.

Shimshi, D., Mayoral, M.L. & Atsmon, D. (1982). Response to water stress in wheat and related wild species. *Crop Science*, **22**, 123–8.

Shpiler, L. & Blum, A. (1986). Differential reaction of wheat cultivars to hot environments. *Euphytica*, **35**, 483–492.

Sobrado, M.A. (1987). Leaf rolling – a visual estimate of water deficit in corn (*Zea mays* L.). *Maydica*, **32**, 9–18.

Sobrado, M.A. & Turner, N.C. (1983). Influence of water deficits on the water relations characteristics and productivity of wild and cultivated sunflowers. *Australian Journal of Plant Physiology*, **10**, 195–203.

Vaadia, Y. (1987). Salt and drought tolerance in plants: regulation of water use efficiency in sensitive and tolerant species. In NATO Conference on *Biochemical and Physiological Mechanisms Associated with Environmental Stress Tolerance in Plants*, University of East Anglia, Norwich, 2–7 August 1987.

Wilcox, J.R., Schapaugh, W.T., Bernard, R.L., Cooper, R.L., Fehr, W.R. & Niehaus, M.H. (1979). Genetic improvement of soybeans in the Midwest. *Crop Science*, **19**, 803–5.

Wit, C.T. de (1958). *Transpiration and Crop Yields*. Wageningen: Centre for Agricultural Publishing and Documentation, Agriculture Research.

Zilberstein, M., Blum, A. & Eyal, Z. (1985). Chemical desiccation as a simulator of postanthesis speckled leaf blotch stress. *Phytopathology*, **75**, 226–30.

A.R. YEO AND T.J. FLOWERS

# 12 Selection for physiological characters – examples from breeding for salt tolerance

## Introduction

The areas of land affected by soil salinity are ill defined, because detailed maps are available for few areas only, and consequently global estimates vary widely (Flowers, Hajibagheri & Clipson, 1986). Figures are more reliable for agricultural land in current or recent useage. On this basis some 2 (out of 15) million km$^2$ of land used for crop production is salt affected, and 30–50% of irrigated land (Flowers et al., 1986). This makes salinity a major limitation to food production; indeed it has been recognised as the largest single soil toxicity problem in tropical Asia (Greenland, 1984). Land suited to intensive agriculture is a finite resource, and it has been estimated that the potential for increase is only some 67%, not much more than projected population increases by the end of the century (see Toenniessen, 1984). Dependence upon marginal land, and the need to reclaim land lost already to salinity, appears inevitable. If it is an agricultural necessity to grow plants in saline environments, then there are three possibilities:

1. *Change the environment.* We are used to solving environmental problems by relatively simple interventions; adding fertilisers, pesticides and water. However, it is much more difficult to remove an excess than it is to supplement a deficiency. A technological solution to the salinity problem has many possibilities but all are costly. In the case of staple foods then either their market value (commercial) or the resources of their consumers (subsistence) precludes such a solution.

2. *Improve the resistance of the crop.* The overwhelming effort in plant breeding is given over towards what may loosely be termed the 'improvement of existing crops'. However, we may be attempting to make some 15 or so species which provide our staple foods (Heywood, 1978) grow in a range of environments for which natural selection has resulted in thousands of different species. 'Improving' a crop in relation to an environmental stress means extending its distribution far beyond its natural range. The

enormous successes that man has had so far with the domestication and adaptation of crop plants should not lull us into ignoring the facts that there may be environments to which some of these species just cannot be adapted.

3. *Change the crop*. Changing crops means growing either a different (and more resistant) crop, or the domestication of new species. There are logical and logistic arguments that it is simpler or easier to domesticate new species which are already salt tolerant than to turn species in current usage into something they are not. Against this are sociological pressures by the consumer who wants to (and often has the belief that science and technology can) preserve the *status quo*.

This paper concentrates on option (2), the modification of existing crops. The other options are assessed in more detail by Epstein *et al.* (1980), Malcolm (1983) and O'Leary (1984).

### Range of salt resistance and tolerance

Crop species show a spectrum of responses to salt, although all have their growth (and yield) reduced by salt (Maas & Hoffman, 1976; Table 1); in contrast, some of the natural flora (halophytes) grow maximally at salt concentrations which would be toxic to other species. The response of crops to salt is blurred by seasonal and climatic variation, related to the known interactions between salt uptake and transpiration as affected by light, temperature and saturation vapour pressure deficit. Growth in salt-affected fields is uneven due to their heterogeneity; areas of high salinity or alkalinity existing in immediate proximity to areas of low salinity (Malcolm, 1983 and references therein; Richards, 1983). The situation may be further complicated since resistance alters according to the criterion of plant productivity used to define it (survival, vegetative growth, grain yield: see Flowers & Yeo, 1989), making comparisons between species, and in particular between crops and the natural flora, intensely difficult. Salinity resistance, like yield, is an integral value with many components, not a clear determinate property.

In spite of these difficulties the available data on crop performance (see Table 1) shows that the external concentration at which yield reduction becomes critical varies according to the crop. The approach to improving the salt resistance depends upon the external concentration at which improvement is sought.

Table 1. *Effect of soil salinity of the yield of various crops*[a]

| Category | Crop | EC[b] causing 50% yield reduction |
|---|---|---|
| Grain | Corn (*Zea mays*) | 5.8 |
| | Rice | 7.0 |
| | Wheat | 13.0 |
| | Barley | 18.0 |
| Vegetable | Bean | 3.7 |
| | Spinach | 8.5 |
| Forage grasses | Wild rye | 10.8 |
| | Tall wheatgrasses | 19.5 |
| Other | Soybean | 7.5 |
| | Sugar beet | 15.5 |
| | Cotton | 17.5 |

[a]Data from Maas & Hoffman (1976).
[b]Electrical conductivity (dS m$^{-1}$).

1. *Low external concentration of salts.* Under these conditions osmotic adjustment to the *external* salt concentration would be metabolically acceptable, even if produced exclusively with NaCl and even if there was poor compartmentation of this salt between vacuole and cytoplasm. Salt damage at low salinity is due largely to excessive salt entry, exacerbated by the distribution of salt within the tissue. Restriction of salt entry alone (provided this was not confounded by intracellular distribution) would be enough to increase resistance. Species like *Oryza sativa* and *Zea mays*, which may be damaged seriously, even killed, at external concentrations as low as 50 mol m$^{-3}$, are examples. The largest *relative* (though not absolute) improvements are likely to come here. There is in most cases the alternative of changing crops to another existing crop which is more tolerant, though other conditions may preclude this. For instance, rice, though salt sensitive, is very tolerant to flooding; it could not be replaced by a more salt-resistant crop which would suffer from waterlogging and submergence in the rainy season.

2. *Moderate soil salt concentration.* These are external concentrations at which substantial osmotic adjustment in the plant is required, and could not be achieved with NaCl without compartmentation between cytoplasm and vacuole. Damage may still be attributable to excessive internal concentrations, but simply minimising salt uptake is not a sufficient answer. The concentration of osmotically active solutes within the cells has to be

adequate to adjust to the low external water potential without producing excessive cytoplasmic $Na^+ + K^+$, $Cl^-$, or $Na^+:K^+$, without amplifying the problems by introducing localised water deficit from apoplastic salt loads, and without excessive dependence on vacuolar adjustment with organic solutes. The latter is undesirable because it will divert a substantial proportion of the resources from the usual agricultural aim, which is seed production. *Hordeum* and *Triticum* would fall into this category.

The potential for relative increase in resistance is likely to have a ceiling imposed by glycophyte physiology in terms of intracellular compartmentation and the lack of facilities (succulence or glands) to moderate leaf salt concentration. Trades-off between cost and benefit (between yield and resistance), though not inevitable, are likely. The crop cannot be replaced with a more resistant one as these are the most resistant of current crops, unless this is by domesticating an halophyte.

3. *High external salinity.* These are external concentrations at which osmotic adjustment cannot be achieved with inorganic solutes without substantial and sustained intracellular compartmentation of salt between cytoplasm and vacuole, and at which either succulence or salt glands are essential to provide for regulation of leaf salt concentrations in fast-growing species. Plants growing well in these conditions are by definition halophytes. Those which survive with very low growth rates and high tissue organic solute concentrations (usually lacking both glands and succulence) are of ecological interest but unlikely to be of agricultural value.

Halophytes can be grown on land which is otherwise agriculturally worthless. The agricultural and economic potential of halophytes has been established, though none is of significant direct use as a food source. Changing the characteristics of natural halophytes to increase their agricultural worth (domestication) is an alternative (or a parallel) to increasing the salt tolerance of non-halophytes. If developed, such crops could be more productive on moderately saline land than resistant forms of existing glycophytic crops.

### Role of selection for physiological characteristics

The potential sites of salt damage and interference are manifold, and at three levels.

1. *Soil structure.* High salt concentrations, and high sodium adsorption ratios in particular, adversely affect the physical properties of the soil (Davidson & Quirk, 1961), altering such parameters as particle size and hydraulic conductance.

2. *Soil/root interactions.* High external concentrations make the acquisition of water and nutrients difficult because of the low water potential of the soil solution, and of chemical competition between saline and nutrient ions.

3. *Within the plant.* Excessive concentrations of some ions occur in the tissue overall, in the cytoplasm, or in the apoplast. Effects include metabolic inhibition, interference with protein synthesis, cellular dehydration, stomatal closure and early senescence of leaves. Since both cytoplasm and apoplast are small compartments, imbalance between compartments may amplify the effects of excess salt, resulting in toxicity despite apparently moderate overall tissue concentrations.

It is very unlikely that there will be a single adaptation which renders the plant able to cope with an environmental constraint which affects, in one way or another, almost every plant process. Even in dealing with salt exclusion in isolation, we are concerned immediately with (1) the relative affinities of transport proteins and ion channels for saline and nutrient ionic species, (2) the chemical composition and structural arrangement of the membranes which affect their passive permeability to saline ions, (3) the ontogeny and architecture of the root system which affects the size of the transpirational bypass flow, (4) the membrane potential and cytoplasmic ion concentrations which determine the free energy change of ions entering and leaving the root and (5) the transpirational volume flow which advects salt to the root and is responsible for net transport to the shoot in the xylem. The main argument to be made from the physiological standpoint which affects breeding strategy, is that salt resistance (or tolerance) is not the consequence of any single, simple factor.

We can compare this with, for instance, resistance to a particular insect, which may result from the production of a single chemical, or efficiency in acquiring a particular nutrient, which could result from the tertiary structure of a single species of transport protein. Although oversimplifying, this does serve to contrast the nature of whole-plant/environment interactions (like salt resistance) with other characters with which we have come to expect success in plant breeding. Another point whose importance will grow in the future as techniques for genetic manipulation develop is that, unlike the insecticidal chemical or transport protein, a great deal of effort will be needed to recognise any of the gene products of salt-resistance mechanisms.

We are, then, faced with breeding to improve something which consists of a multiplicity of contributing factors. The overall phenotypic performance of a plant may thus be a poor indicator of its genetic value in improvement, since useful attributes can be masked by other factors resulting in overall sensitivity. Screening (applying an artificial selection pressure and looking for well-adapted plants) can only detect existing *phenotypic* variation. The absence of phenotypic variation does not mean the absence of useful genotypic variation, however. Selection pressure is needed to collect genetic traits into a single genotype, and there has been rather little natural selection for salt resistance in many crop species. Thus the underlying principle of a screening trial, *that good genotypes will be accepted and only bad genotypes will be rejected on the basis of their phenotypes*, is not necessarily applicable to salinity resistance. It is thus necessary to identify and recognise the functioning of physiological processes which constitute characteristics which assist in the resistance of salinity, in situations in which the phenotype as a whole is not distinctive and may even be salt sensitive. This may serve a number of breeding objectives, illustrated by the examples which follow.

### Examples
#### Rice
Despite vast efforts over 20 years, rice remains one of the most salt-sensitive of crops. Efforts have centred on screening the existing germplasm, and the more resistant varieties have proved to be older, traditional (and therefore non-dwarfed) land races. Although some of their resistance has been incorporated into modern varieties, these have not increased in resistance beyond that of their parents, and do not achieve the resistance of barley or wheat (Ponnamperuma, 1984).

Since rice is affected by salinities as low as 50 mol m$^{-3}$, but has salt concentrations greater than this in the tissue, we might judge the osmotic adjustment would be at least adequate. We might then expect that damage was attributable to excessive concentrations of salt in the leaves, and thus that survival would be correlated inversely with salt content. This is broadly true, but the correlation accounts for only some 30% of the variation, rather less than is accounted for by plant vigour alone. This leads to two conclusions; that there is substantial and useful variation in Na uptake in rice and that, although uptake is part, it is far from being the whole story (Yeo & Flowers, 1986). We now recognise a number of physiological characteristics which mitigate, or modify, the consequences of salt uptake (Table 2).

Salt damage is due to the product of xylem concentration and transpiration rate over the life of the leaf. There are several pathways for salt entry into the root. In spite of many reports in the literature of correlative associations between the lipid analyses of relatively crude root membrane preparations and salt resistance (see Kuiper, 1985), we have not been able to demonstrate any significant differences in the microsomal membranes of 24 rice lines covering a wide range of salt resistance (Fig. 1). Direct apoplastic contact in regions of the root where the permeability barrier of the endodermis is ineffective (the bypass flow) is, however, important in rice (Yeo, Yeo & Flowers, 1987). Even so, rice varieties already show at least 95% exclusion (based on comparing xylem and external $Na^+$ concentrations: see Yeo *et al.*, 1987; Flowers, Salama & Yeo, 1988) and this may be difficult to improve upon. Features which mitigate against the adverse consequences of this inevitable input are as important as initial entry, may have more capacity for improvement, and may be easier to alter.

Table 2. *Some physiological characters which affect salt resistance in rice*

Apoplastic transport of sodium across the root
Plant vigour (degree of dwarfing)
Water-use efficiency
Leaf compartmentation
Tissue-tolerance (apoplast/protoplast balance)

Fig. 2. The effect of salinisation (NaCl, 50 mol m$^{-3}$ at 14 days) on survival of leaves of rice plants. Open circles (salinised), closed circles (non-salinised), leaf 2 (leaf 1 is the oldest) to the left of the figure and leaf 4 to the right. The shaded area represents lost leaf life due to salinity. Data of P. Izard, T.J. Flowers & A.R. Yeo (unpublished).

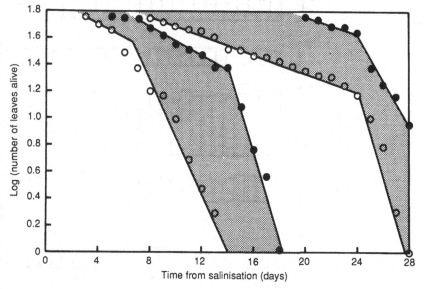

Fig. 1. Relative composition of root microsomal membranes from 24 land races, varieties and breeding lines of rice which differ in their salt resistance. Campesterol, Stigmasterol and Sitosterol as % of total sterols; 16:0, 18:1, 18:2 and 18:3 fatty acids as % of total fatty acids; Na transport on a relative scale from (1) lowest to (9) highest. Data of D.R. Lachno, T.J. Flowers & A.R. Yeo (unpublished).

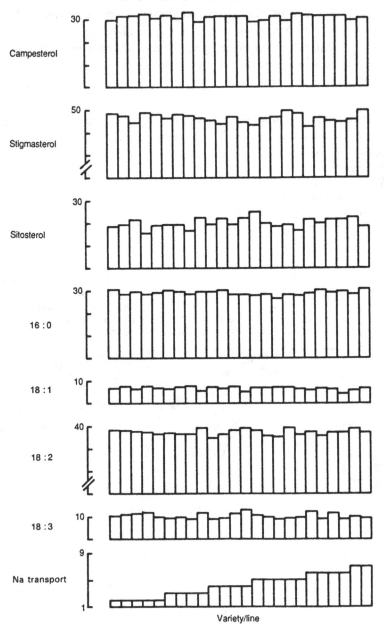

Table 3. *Physiological characteristics of some varieties and lines of rice.*

Relative performance (on a scale from 1 (good) to 9 (bad)) in $Na^+$ transport, tissue-tolerance, leaf compartmentation and vigour

| Designation | $Na^+$ transport | Tissue-tolerance | Leaf compartmentation | Vigour |
|---|---|---|---|---|
| IR4630-22-2-5-1-2 | 2 | 1 | 4 | 7 |
| IR15324-117-3-2-2 | 9 | 7 | 5 | 6 |
| IR10167-129-3-4 | 6 | 2 | 3 | 8 |
| Nona Bokra | 1 | 7 | 6 | 2 |

High vigour provides more shoot tissue in which to distribute the salt which is transported and, for a given rate of net transport, the average tissue concentration will be lower in the variety with the higher growth rate (Yeo & Flowers, 1984). All else being equal, salt concentration in the leaves will be proportional to transpiration per unit growth. More efficient use of water will reduce salinity damage (Flowers *et al.*, 1988). It is an important observation at this stage (because it so affects how parental lines are chosen) that characters which increase salt resistance may do so for purely incidental reasons. Water-use efficiency is an example: it is useful in resisting salinity, but it most likely evolved as a response to drought; a drought-resistant parent might, consequently, be a useful source of genetic material in a programme aimed at increasing salinity resistance.

Even a moderate quantity of salt reaching the leaves has a drastic effect on photosynthesis and leaf ultrastructure, much more than could be accounted for by the *average* tissue concentration (Flowers *et al.*, 1985). Salt may accumulate in the apoplast (because it is not taken up fast enough by the cells of the leaf), and this would result in severe localised water deficit (Oertli, 1968). Differences in apoplast/protoplast balance are thought to be responsible for varietal differences in tissue salt load which can be accommodated (tissue tolerance; Yeo & Flowers, 1986). The xylem concentration of $Na^+$ is very much lower to young leaves than to older leaves (Yeo *et al.*, 1985). This is advantageous in salt resistance because it means that at least some leaves are protected from salt, which otherwise causes premature leaf death (Yeo & Flowers, 1984; Fig. 2).

Varietal differences exist in all of these characteristics, and it is our theory that parents conferring the required traits should be combined. Depending on heritability and combining ability, we may add together genes for a number of useful characteristics, and a variety more resistant than any which has arisen by chance should be produced. Salt resistance is thus treated as a complex whole-plant problem, not a single factor. Lines which are good for one character are commonly poor in others (Table 3), and may have low overall phenotypic resistance. Thus this procedure contrasts strongly with phenotype screening, which will favour a variety which is mediocre at everything over one that is excellent in one, but poor in others.

### Tomato

The tomato (*Lycopersicon esculentum* Mill.) is an important horticultural crop that is grown in glasshouses in temperate regions, but is field grown particularly in areas with a Mediterranean climate (Davies & Hobson, 1981). These fields are frequently subject to salinity and there has been an interest in breeding for salt tolerance for about 20 years. Within the genus *Lycopersicon*, there are some seven species which can all be hybridised, thus raising the possibility of transferring genes from wild relatives to the domesticated *L. esculentum*. For salt tolerance this is feasible, in principle, since the wild relatives *L. peruvianum*, *Solanum pennellii* and *L. cheesmanii* are more tolerant than *L. esculentum* (Tal, 1971; Rush & Epstein, 1976). These wild relatives behave as halophytes as far as sodium accumulation under saline conditions is concerned, accumulating higher concentrations in their leaves than does *L. esculentum* (Rush & Epstein, 1976; Tal & Shannon, 1983). This character has been used to increase the salt tolerance of an edible tomato by hybridising *L. esculentum* with *L. cheesmanii* and making successive backcrosses with the cultivated variety, Walter, as the recurrent seed parent (Rush & Epstein, 1981). The

Fig. 3. Fruit yields at various salinities (FW, freshwater; SW, seawater) of *Lycopersicon esculentum* cv. Walter, and the $F_2$ and backcross ($F_3BC_1$) resulting from its hybridisation with *L. cheesmanii*. Data of Rush & Epstein (1981).

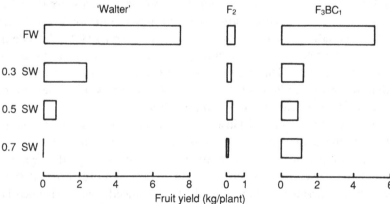

yield of Walter was greater than that of the hybrid under non- and low-saline conditions, but the hybrid continued to yield under conditions where Walter failed. Although the yield per plant of the hybrid was low, later backcrosses produced increased yields at the higher salinities (Fig. 3). This suggests that genetic information relating to the resistance of the wild relative has been transferred to the domestic genome and, in addition, that successive backcrossing can restore yielding ability.

### Wheat

*Thinopyrum bessarabicum* is a perennial species within the graminaceous tribe Triticeae, which is more salt resistant than the annual *Triticum* species including the bread wheats. It has been hybridised with a cultivar of *Triticum aestivum* (Forster & Miller, 1985) and an amphidiploid produced by colchicine treating this hybrid has been investigated by Gorham *et al.* (1986). The amphidiploid was more resistant, in terms of survival and ability to produce grain at moderate (250 mol m$^{-3}$) external salinity, than either the parental cultivar (Chinese Spring) or a variety (Kharchia) with a field reputation for resistance to salinity. The greater resistance of the amphidiploid was attributed to its inheritance of more efficient exclusion of Na$^+$ and Cl$^-$ from the younger leaves and reproductive tissue (Gorham *et al.*, 1986), thus behaving in a similar way to the rice case described earlier.

Restoring agricultural worth after amalgamating a complete alien genome is the inevitable counterpart to hybridisation with a wild relative. The problem facing a breeder is to make a domestic crop from the unimproved hybrid. In this case it expresses much of the *Thinopyrum* genome which has many agriculturally undesirable qualities (Gorham *et al.*, 1986). The advantages of transferring only the relevant part of the genome would be in drastically reducing the amount of subsequent backcrossing or selection. Wheat also provides an example of the localisation of a physiological character important in salt resistance, Na$^+$/K$^+$ selectivity (Wyn Jones, Gorham & McDonnell, 1984), to a single chromosome (Gorham *et al.*, 1987). The hexaploid bread wheats have a combination of three ancestral genomes, designated A, B and D, the D genome being derived from *Aegilops squarrosa*. This species, and some of the AABBDD hexaploids, express high K$^+$/Na$^+$ selectivity compared with the AABB tetraploids. This suggests that the character for K$^+$/Na$^+$ selectivity derives from the D genome (Shah *et al.*, 1987). A complete set of disomic substitution lines, in which one D genome chromosome replaces either the A or B genome homeologues in an AABB tetraploid, have been examined for Na$^+$ and K$^+$ uptake in saline conditions. The results showed (Fig. 4) that K$^+$/Na$^+$ selectivity was associated with chromosome 4 of the D genome (Gorham *et al.*, 1987).

*Cell culture*

With the establishment of protocols allowing the regeneration of plants from single cells and protoplasts, the possibility exists for selecting cells which show salt resistance, and regenerating salt-resistant plants from them (see Rains, Chapter 10). The selection of cell lines which are apparently salt resistant has proved to be easy. In relatively few cases has the procedure been taken through to assess whether the regenerated plants are salt resistant. Unless this is done, it is impossible to establish the difference between a salt-resistant cell line which might carry useful attributes, one that is osmotically (but not especially salt) resistant, and one that relies on a mechanism made feasible only by the abundant resources supplied in the nutrient medium. It is not even possible to establish clearly whether the difference in behaviour is genetic in origin.

Fig. 4. K/Na ratios and (Na + K) concentrations in sap from leaves of disomic D genome substitution lines in *Triticum turgidum* cv. Langdon grown in NaCl (150 mol m$^{-3}$ plus CaCl$_2$ at 7.5 mol m$^{-3}$). Data of Gorham *et al.* (1987)

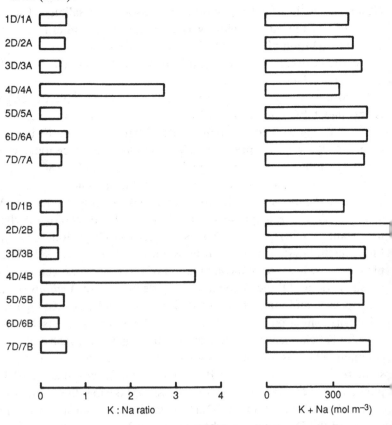

Fig. 5. Effect of salinity on the growth (as freshweight increase) of whole plants and callus of a sensitive glycophyte (*Phaseolus*), a resistant glyco-phyte (*Beta*) and of two halophytes (*Atriplex* and *Suaeda*). Data of Smith & McComb (1981).

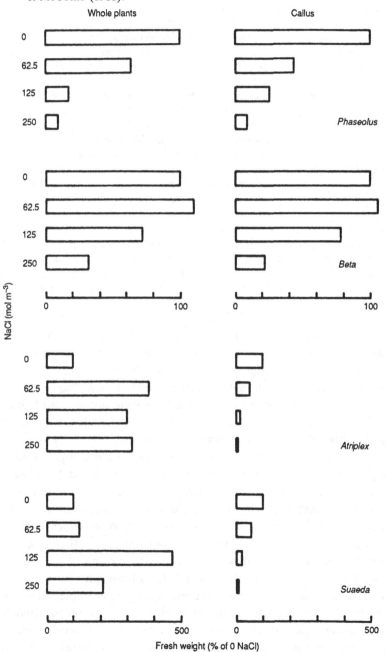

Table 4. *Salt resistance of plants regenerated from cell lines selected for tolerance of salt in vitro*

| Species | Remarks | Ref. |
|---------|---------|------|
| *Colocasia esculenta* | Not tested; regenerated plants died | (1) |
| *Linum usitatissimum* | Increased; measured plant height | (2) |
| *Medicago sativa* | Not fully determined. Lethal genetic changes after long-term culture | (3) |
| *Nicotiana tabacum* | Increased; determined survival in saline irrigation. Phenotype persistence dependent on, or enhanced by, presence of NaCl | (4) |
| | Increased; determined by growth on salinised medium. Selected plants were hexaploid | (5) |
| *Pennisetum purpureum* | Not tested | (6) |
| | No increase; preliminary statement, but no data | (7) |
| *Sorghum bicolor* | No difference from parental line; measured seedling height | (8) |

*Sources:* (1) Nyman, Gonzalez & Arditti (1983); (2) McHughen & Swartz (1984); (3) T.P. Croughan, PhD thesis, University of California, Davis, cited by Stavarek & Rains (1984); (4) Nabors *et al.* (1980); (5) Bressan *et al.* (1985); (6) Bajaj & Gupta (1986); (7) Chandler & Vasil (1984); (8) Bhaskaran, Smith & Schertz (1986).

The overwhelming advantages of an approach to breeding based on cell culture are the large numbers of individual cells which can be handled, and the ease of screening. There is also the potential for a higher frequency of genetic variation than is found in intact plants (Larkin & Skowcroft, 1981). However, there are problems of relating *in vivo* and *in vitro* behaviour which are illustrated by the data of Smith & McComb (1981) shown in Fig. 5. Although the data are complicated somewhat by the choice of fresh weight as the growth parameter, it is clear that there is a disparity between the growth response of the halophytes (*Atriplex* and *Suaeda*) as plants and as callus to salt. While the correlation between *in vivo* and *in vitro* behaviour is much better in *Beta* and *Phaseolus*, this can not be taken as a generalisation. There may be salt-resistance mechanisms which are based at the cellular level and observed at the plant level, but other mechanisms depend upon the functional integrity of the whole plant.

Success in regenerating salt-resistant plants from salt-resistant cells has been limited (Table 4). The most extensively quoted example is the work of Nabors *et al.* (1980) who found that tobacco plants regenerated from NaCl-resistant cell lines survived salinity better than did plants regenerated from unselected lines. The authors themselves were justifiably cautious, reporting that the presence of NaCl during regeneration and subsequent generations was important in the phenotypic persistence of the trait. Nabors *et al.* (1980) suggest gene amplification (a potential contributor to somaclonal variation; Larkin & Skowcroft, 1981) as one possible explanation. Plants regenerated by Bressan *et al.* (1985), which also expressed increased resistance, proved to be hexaploid. While this does constitute selection of a trait expressed at cell and plant level, it is the general quality of vigour, rather than a mechanism of salt resistance *per se*, which is selected.

There is much unexplored scope for physiological analysis of resistant lines to see how they do differ from the unselected populations. A line which maintained its cell volume, had high internal $K^+$ and a compatible solute would have obvious agricultural potential: a line that survived by drastic reduction in cell volume and heavy reliance on supplied organic solutes would have little to offer in a practical context.

### Concluding remarks

Selection on the basis of physiological characters has several advantages in breeding for complex whole-plant responses such as the resistance to salinity. Where resistance depends on the summation of several characters, each may be phenotypically cryptic because it is not alone sufficient to confer salt resistance; these characters need to be recognised in their own right. Some characters may contribute to resistance incidentally and, once recognised, the best parents may come from sources unrelated to salinity, as in the case of water use efficiency. Understanding the mechanism of salt resistance in a particular species (and this differs drastically from species to species) provides insight on which to select parental species in wide crossing, and indicates quantitative characteristics (like $K^+/Na^+$ selectivity) which may be localised to some part of the genome. Cell-based selection methods could contribute to breeding for resistance if the physiological basis of cell resistance can be identified as one which may be expressed in regenerated plants.

### References
Bajaj, Y.P.S. & Gupta, R.K. (1986). Different tolerance of callus cultures of *Pennisetum americanum* L. and *P. purpureum* Shum. to sodium chloride. *Journal of Plant Physiology*, **125**, 491–5.
Bhaskaran, S., Smith, R.H. & Schertz, K.F. (1986). Progeny screening of sorghum plants regenerated from sodium chloride-selected callus for salt tolerance. *Journal of Plant Physiology*, **122**, 205–10.
Bressan, R.A., Singh, N.K., Handa, A.K., Kononowicz, A. & Hasegawa, P.M. (1985). Stable and unstable tolerance to NaCl in cultured tobacco cells. In *Plant Genetics*, ed. M. Freeling, pp. 755–69. New York: Alan R. Liss.
Chandler, S.F. & Vasil, I.K. (1984). Selection and characterisation of NaCl tolerant

lines from embryogenic cultures of *Pennisetum purpureum* Schum. (Napier grass). *Plant Science Letters*, **37**, 157–64.

Davidson, J.L. & Quirk, J.P. (1961). The influence of dissolved gypsum and pasture establishment on irrigated sodic clays. *Australian Journal of Agricultural Research*, **12**, 100–10.

Davies, J.N. & Hobson, G.E. (1981). The constituents of tomato fruit; the influence of environment, nutrition and genotype. *CRC Critical Reviews in Food Science and Nutrition*, pp. 205–80.

Epstein, E., Norlyn, J.D., Rush, D.W., Kingsbury, R.W., Kelley, D.B., Cunningham, G.A. & Wrona, A.F. (1980). Saline culture of crops: a genetic approach. *Science*, **210**, 399–404.

Flowers, T.J. & Yeo, A.R. (1989). Effects of salinity on plant growth and crop yields. In *Biochemical and Physiological Mechanisms Associated with Environmental Stress Tolerance in Plants*, ed. J. Cherry. Berlin: Springer-Verlag, (in press).

Flowers, T.J., Duque, E., Hajibagheri, M.A., McGonigle, T.P. & Yeo, A.R. (1985). The effect of salinity on the ultrastructure and net photosynthesis of two varieties of rice: further evidence for a cellular component of salt resistance. *New Phytologist*, **100**, 37–43.

Flowers, T.J., Hajibagheri, M.A. & Clipson, N.C.W. (1986). Halophytes. *Quarterly Review of Biology*, **61**, 313–37.

Flowers, T.J., Salama, F.M. & Yeo, A.R. (1988). Water use efficiency in rice (*Oryza sativa* L.) in relation to resistance to salinity. *Plant, Cell and Environment* **11**, 453–9.

Forster, B.P. & Miller, T.E. (1985). 5B deficient hybrid between *Triticum aestivum* and *Agropyron junceum*. *Cereal Research Communications*, **13**, 93–5.

Gorham, J., Forster, B.P., Budrewicz, E., Wyn Jones, R.G., Miller, T.E. & Law, C.N. (1986). Salt tolerance in the Triticeae: solute accumulation and distribution in an amphidiploid derived from *Triticum aestivum* cv. Chinese Spring and *Thinopyrum bessarabicum*. *Journal of Experimental Botany*, **37**, 1435–49.

Gorham, J., Hardy, C., Wyn Jones, R.G., Joppa, L. & Law, C.N. (1987). Chromosomal location of a K/Na discrimination character in the D genome of wheat. *Theoretical and Applied Genetics*, **74**, 484–8.

Greenland, D.J. (1984). Exploited plants: rice. *Biologist*, **31**, 291–25.

Heywood, V.H. (ed.) (1978). *Flowering Plants of the World*. Oxford: Oxford University Press.

Kuiper, P.J.C. (1985). Environmental changes and lipid metabolism of higher plants. *Physiologia Plantarum*, **64**, 118–22.

Larkin, P.J. & Skowcroft, W.R. (1981). Somaclonal variation – a novel source of variability from cell cultures for plant improvement. *Theoretical and Applied Genetics*, **60**, 197–214.

Maas, E.V. & Hoffman, G.J. (1976). Crop salt tolerance: evaluation of existing data. In *Managing Water for Saline Irrigation*, ed. H.E. Dregre. Lubbock: Texas Technical University.

Malcolm, C.V. (1983). Wheatbelt salinity, a review of the salt land problem in South-Western Australia. *Technical Bulletin No. 52*. Perth: Western Australia Department of Agriculture.

McHughen, A. & Swartz, M. (1984). A tissue-culture derived salt-tolerant line of flax (*Linum usitatissimum*). *Journal of Plant Physiology*, **117**, 109–17.

Nabors, M.W., Gibbs, S.E., Bernstein, C.S. & Meis, M.M. (1980). NaCl-tolerant tobacco plants from cultured cells. *Zeitschrift für Pflanzenphysiologie*, **97**, 13–17.

Nyman, L.P., Gonzalez, C.J. & Arditti, J. (1983). *In vitro* selection for salt tolerance of taro (*Colochasia esculenta* var *antiquorum*). *Annals of Botany*, **51**, 229–36.

Oertli, J.J. (1968). Extracellular salt accumulation, a possible mechanism of salt injury in plants. *Agrochimica*, **12**, 461–9.

O'Leary, J.W. (1984). The role of halophytes in irrigated agriculture. In *Salinity Tolerance in Plants: Strategies for Crop Improvement*, ed. R.C. Staples and G.A. Toenniessen, pp. 285–300. New York: John Wiley.

Ponnamperuma, F.N. (1984). Role of cultivar tolerance in increasing rice production on saline lands. In *Salinity Tolerance in Plants: Strategies for Crop Improvement*, ed. R.C. Staples and G.A. Toenniessen, pp. 255–71. New York: John Wiley.

Richards, R.A. (1983). Should selection for yield in saline regions be made on saline or non-saline soils? *Euphytica*, **32**, 431–8.

Rush, D.W. & Epstein, E. (1976). Genotypic responses to salinity. Differences between salt-sensitive and salt-tolerant genotypes of the tomato. *Plant Physiology*, **57**, 162–6.

Rush, D.W. & Epstein, E. (1981). Breeding and selection for salt tolerance by the incorporation of wild germplasm into a domestic tomato. *Journal of the American Society for Horticultural Science*, **106**, 699–704.

Shah, S.P., Gorham, J., Forster, B.P. & Wyn Jones, R.G. (1987). Salt tolerance in the Triticeae: the contribution of the D genome to cation selectivity in hexaploid wheat. *Journal of Experimental Botany*, **37**, 254–69.

Smith, M.K. & McComb, J.A. (1981). Effect of NaCl on the growth of whole plants and their corresponding callus cultures. *Australian Journal of Plant Physiology*, **8**, 267–75.

Stavarek, S.J. & Rains, D.W. (1984). Cell culture techniques: selection and physiological studies of salt tolerance. In *Salinity Tolerance in Plants: Strategies for Crop Improvement*, ed. R.C. Staples and G.A. Toenniessen, pp. 321–34. New York: John Wiley.

Tal, M. (1971). Salt tolerance in the wild relatives of the cultivated tomato: responses of *Lycopersicon esculentum*, *L. peruvianum* and *L. esculentum minor* to sodium chloride solution. *Australian Journal of Agricultural Research*, **22**, 631–8.

Tal, M. & Shannon, M.C. (1983). Salt tolerance in the wild relatives of the cultivated tomato: responses of *Lycopersicon esculentum*, *L. cheesmanii*, *L. peruvianum*, *Solanum pennellii* and F₁ hybrids to high salinity. *Australian Journal of Plant Physiology*, **10**, 109–17.

Toenniessen, G.A. (1984). Review of the world food situation and the role of salt-tolerant plants. In *Salinity Tolerance in Plants: Strategies for Crop Improvement*, ed. R.C. Staples and G.A. Toenniessen, pp. 399–413. New York: John Wiley.

Wyn Jones, R.G., Gorham, J. & McDonnell, E. (1984). Organic and inorganic solutes as selection criteria for salt tolerance in the Triticeae. In *Salinity Tolerance in Plants: Strategies for Crop Improvement*, ed. R.C. Staples and G.A. Toenniessen, pp. 189–203. New York: John Wiley.

Yeo, A.R. & Flowers, T.J. (1984). Mechanisms of salinity resistance in rice and their role as physiological criteria in plant breeding. In *Salinity Tolerance in*

*Plants: Strategies for Crop Improvement*, ed. R.C. Staples and G.A. Toenniessen, pp. 151–70. New York: John Wiley.

Yeo, A.R. & Flowers, T.J. (1986). The physiology of salinity resistance in rice (*Oryza sativa* L.) and a pyramiding approach to breeding varieties for saline soils. *Australian Journal of Plant Physiology*, **13**, 161–73.

Yeo, A.R., Yeo, M.E., Caporn, S.J.M., Lachno, D.R. & Flowers, T.J. (1985). The use of $^{14}$C-ethane diol as a quantitative tracer for the transpirational volume flow of water, and an investigation of the effects of salinity upon transpiration, net sodium accumulation and endogenous ABA in individual leaves of *Oryza sativa* L. *Journal of Experimental Botany*, **36**, 1099–109.

Yeo, A.R., Yeo, M.E. & Flowers, T.J. (1987). The contribution of an apoplastic pathway to sodium uptake by rice roots in saline conditions. *Journal of Experimental Botany*, **38**, 1141–53.

R.B. AUSTIN

# 13 Prospects for improving crop production in stressful environments

## Introduction

In recent years the term 'stress' has been used in an uncritical and non-specific way by administrators of agricultural and plant research, and by some plant physiologists, plant breeders, and molecular biologists. As a consequence, the view is often propagated that crop varieties may 'soon' be produced which will be much more stress resistant or stress tolerant than those presently grown. In this concluding chapter, I shall urge caution over this optimistic view before drawing attention to objectives which it may be possible to achieve.

### Stress – a normal condition of plants

In common with all living organisms, plants have to acquire the materials from which they are made from their surroundings. In the case of plants this acquisition usually has to be made against a concentration gradient. For land plants, tissue water has to be retained as well as acquired against gradients of water potential which can vary from small, as in humid tropical forests, to very large, as in hot, dry deserts. The concentrations of many essential ions in plants are often a thousand-fold greater than those in the soil solution from which their roots extract these ions. In the sense that plants have to expend metabolic energy to acquire their resources and invest in structures to acquire and conserve them in the face of limited supply, stress is a normal condition of plants, though of variable severity. Commonly, plants are considered to be under stress when they experience a relatively severe shortage of an essential constituent or an excess of a potentially toxic or damaging substance. This is the context for the discussion of what is usually meant when physiologists, ecologists and agriculturalists talk about stress (see Chapter 1).

### The diversity of stress

Stressful environments are often characterised by the occurrence of more than one stress simultaneously and where this is so the means of overcoming the stresses, by breeding or management, may similarly be complicated. Thus salinity is often associated with drought, or with water-logging and/or poor soil structure, and drought with high temperature. Low pH may induce manganese toxicity, aluminium toxicity or calcium deficiency depending on the soil. Numerous other examples could be listed. Typically, stress varies with time, both in kind and intensity. This variability occurs on different time scales – from hour to hour, day to day, week to week and season to season.

These generalisations about stress need to be borne in mind by physiologists when studying stress, by agronomists when they seek to devise means of alleviating stress by appropriate management of crops, and by plant breeders, geneticists, biochemists and molecular biologists when engineering for genetic resistance.

### The productivity of different ecosystems

The productivity of different ecosystems can be taken as an indication of the productivity that may be expected from crops which are optimally adapted to given environments. Data originally assembled as part of the International Biological Programme are summarised in Table 1, which also gives for comparison the range of productivity of cereal crops and sugar cane. Clearly, where there are limitations imposed by the length of the season and by the availability of water (or nutrients in the case of ocean plankton), productivity is lower. For example, productivity of grasslands is lower than for wet tropical forests, littoral seaweeds or sugar cane. Thus it is totally unrealistic to expect that the productivity of deserts or tundra ecosystems could approach that of wet tropical forests. Interestingly, littoral seaweed ecosystems can be very productive and mangrove swamps moderately productive. This shows that when water is present in abundance, the salinity of sea water does not impose on halophytic species a stress that greatly reduces their productivity. Of course, the productivity of salt deserts is much lower than that of saline wetlands, similar to that of other deserts, and depends on the annual rainfall.

### Managing stress

The best way of avoiding the effects of stress is to prevent the stress from occurring. Where technology and resources permit, this is a primary aim of farmers. They plough and cultivate to improve the soil as a medium for root growth and functioning and avoid the various kinds of stress caused by competition from weeds. They irrigate the soil to avoid severe water deficit or drain it to prevent waterlogging. They control biotic stresses caused by pests and diseases by the application of appropriate agrochemi-

Table 1. *Net primary dry-matter production in different ecosystems,*
$t \, ha^{-1} \, year^{-1}$

| | |
|---|---|
| Wet tropical forests | 30–40 |
| Littoral seaweeds | 10–40 |
| Temperate forests | 10–20 |
| Tropical savannah grassland | 10–20 |
| Temperate saltmarsh (*Spartina*) | 2–10 |
| Temperate grassland | 2–10 |
| Mangrove swamps | 6–8 |
| Tundra | 0.5–2 |
| Ocean plankton | ≈2 |
| Non-polar deserts | 0.1–0.2 |
| Crops: cereals | 1–30 |
| sugar cane | 8–50 |

(Adapted from Cooper, 1975).

cals. Much of crop science is aimed at finding the economic optimum management methods for obtaining maximum yields or maximum profits from crop growing. A very large investment in infrastructure and inputs of fertilisers and crop protection chemicals is needed to enable farmers to obtain maximum yields. What is economic for most farmers in the world falls far short of that necessary to achieve the maximum yields possible given the prevailing regimes of temperature and light. Thus, most crops experience water shortage and so yield less than when fully irrigated. Also there are substantial areas where salinity limits cropping. Although extremes of temperature, deficiencies of essential ions and toxic concentrations of others are now popularly referred to as 'stresses', this article concentrates on drought and salinity.

### Physiological aspects of particular stresses and plant responses
*Drought*
The water relations of plants are reasonably well understood and have been described in numerous text books and reviews (e.g. Monteith, 1973; Paleg & Aspinall, 1981; Lange *et al.*, 1981). The laws of physics determine the behaviour of leaves and, together with the biochemical characteristics of the leaves (Farquhar, von Caemmerer & Berry, 1980), set limits to the assimilation ratio (the ratio of assimilation to transpiration; Cowan & Troughton, 1971; Jones, 1976). Although leaves transmit much of the infra-red radiation from the sun, they absorb strongly in the visible region. To maintain thermal equilibrium this energy has to be dissipated. In well-watered leaves of mesophytes, around 70% of incoming energy is lost

Table 2. *Some features of plants conferring adaptation to water-limited environments*

| Plant feature | Reference |
|---|---|
| 1. Regulation of life cycle timing by photoperiod or temperature so that photosynthesis and transpiration are largely confined to times of the year when water is most freely available and transpirational demand is least. In most crop plants this *phenological* adaptation is the most important and the most easily manipulated feature conferring high yield in water-limited environments. In perennials which grow in arid or semi-arid environments, the deciduous habit is an important means of achieving this objective. | Craufurd *et al.*, 1988 |
| 2. Regulation of life cycle timing so that moisture-sensitive growth stages do not coincide with times when water stress is likely to occur. | Salter & Goode, 1967 |
| 3. Means of reducing radiation load that include small leaves, which may also be reflective and/or hairy, and leaf rolling. Hairy leaves also increase the boundary layer resistance and so augment the stomata as means of restricting water loss. | Nobel, 1983 |
| 4. Thick leaves and/or succulent habit providing large thermal capacity so limiting the increase in temperature of the tissues during the day time. | Gutschick, 1987; Nobel, 1983 |
| 5. Constitutive features of stomatal complexes which give a low stomatal conductance (e.g. 'sunken' stomata, few stomata per unit leaf area). | Jarvis & Mansfield, 1981 |
| 6. Adaptive features of stomata – rapid closure in response to turgor loss (and perhaps other stimuli) – which reduce transpiration when plants become stressed. Alternatively, for ephemerals, maintenance of high stomatal conductance and photosynthetic capacity to permit maximum growth rate during the limited time that water is available. | Mansfield & Davies, 1981; Zeiger, 1983 |
| 7. Ability of tissues to increase their solute concentration (osmoregulate), conferring a temporary advantage by enabling the tissues to maintain turgor at low water potentials by decreasing their osmotic potentials. | Morgan *et al.*, 1986 |
| 8. Metabolic features which increase water use efficiency (a) CAM metabolism (b) $C_4$ metabolism and (c) other $CO_2$ concentrating mechanisms. | Szarek & Ting, 1975 |
| 9. Deep roots, permitting water extraction from moist soil at depth. | Passioura, 1983 |
| 10. Few but long roots to permit a slow but sustainable supply of water to the plant. | Passioura, 1983 |
| 11. Roots with low hydraulic conductivity to permit a slow but sustainable supply of water to the plant. | Passioura, 1983 |

as latent heat of evaporation of water in the process of transpiration, smaller proportions being reflected, re-radiated as long-wave infra-red, and lost as sensible heat by convection, with a minor proportion being used to provide the energy for photosynthesis. In water-stressed leaves of mesophytes these proportions are disturbed because the stomata close and less energy is lost as latent heat. As a consequence, leaves become warmer but, as radiation loss from them is proportional to the fourth power of the temperature, the extent to which they are warmer than their environment is limited. Unfortunately, in these circumstances the temperature of the environment in general increases because of high irradiance. Thus leaves themselves can reach temperatures of 30–40 °C or more. Plants in such environments must therefore be tolerant of high temperature as well as be able to resist or survive water stress. One strategy is to become dormant, a common feature of perennial grasses in Mediterranean climates and trees in hot savannahs. Otherwise plants need to have metabolic adaptation to high temperature, or to have a large heat capacity, features common in succulent plants which grow in hot, dry environments.

The effects of water deficits on plant growth and crop yield have been the subject of a great deal of research. Much is known of individual morphological and anatomical features of plants which enable them to survive and grow as well as possible in water-limited environments (Turner & Kramer, 1980; Srivastava *et al.*, 1987). However, a common feature of many wild species grown in water-limited environments is the possession of a *syndrome* of features which together confer an ability to survive and grow as fast as possible in these conditions. Some of these features are given in Table 2. Together, these features confer drought tolerance. The appropriate combination of features and the level of expression of each feature will depend on species and environment, as the following examples show.

For barley grown in dry Mediterranean environments, recent studies indicate the importance of maximising growth rates during the winter when water availability and water use efficiency is greatest (Cooper *et al.*, 1987). This can be achieved by alleviating phosphorus deficiency, by selecting genotypes which grow rapidly in winter and flower early in summer and have a short grain-filling period. By these means grain yields can be raised substantially, but they are still much lower than those obtained from barley crops grown in the more favourable environments of north-west Europe. Thus, the improved varieties do not grow better during the summer drought than their predecessors, they avoid it. This is a common feature of other crops grown in other dry environments, though it must be emphasised that it is not the only feature conferring drought tolerance.

In the semi-arid tropics where millets (*Pennisetum americanum, Digitaria* and *Eleusine* spp.) are grown, the rainy season varies in date of onset and duration and varieties have evolved which are well adapted to exploit fully the short growing season. Millets grow rapidly during the short rainy season in these areas and grain filling is complete soon after the onset of the dry season. They can also resist short periods of drought during the rainy season, suffering little loss in yield. Of course, as with barley in Mediterranean environments, there is much scope for better crop management (apart from irrigation, for which water is not generally available) which would improve crop yields.

Most features listed in Table 2 would confer drought resistance or survival but would not necessarily increase productivity in arid or semi-arid environments. In temperate climates, where crops often experience moderate drought for short periods, adaptive responses might be exploited to limit the damage which drought could cause. Such features may include modified stomatal responses to stress (Quarrie, 1982), increased ability to osmoregulate (Morgan, Hare & Fletcher, 1986) and/or accumulation of organic solutes such as proline (Paleg & Aspinall, 1981; but see Hanson *et al.*, 1979) and glycine betaine (Grummet, Albrechtsen & Hanson, 1987), or inducible Crassulacean acid metabolism. While the numerous plant features that confer fitness in a given drought-prone environment are all under genetic control, it is likely that within a species no single feature other than life-cycle timing has a major effect on drought tolerance. Therefore, the manipulation of single genes will generally have only modest effects and the benefits from incorporating or enhancing the features individually are likely to give only modest improvements in productivity, water-use efficiency and yield.

From a physiological viewpoint there are intriguing differences among species in drought tolerance. Taking cereals as an example, maize is much more susceptible to drought than sorghum which is more susceptible than pearl millet, and these differences do not appear to be entirely due to the length of the life cycle. A comparative study of these species to identify features responsible for the differences in drought susceptibility might reveal targets for genetic engineering.

Rice is at the other end of the drought-tolerance spectrum from barley, paddy rice being more drought susceptible than upland (rainfed) rice. For much of the time since rice was domesticated, farmers have selected forms for cultivation in rainfed (as distinct from flooded or paddy) conditions. Despite this, upland rice remains very susceptible to drought in comparison with other cereals. A major reason for its drought susceptibility seems to be that it is unable to regulate its transpirational water loss as effectively as other cereals. As a result, when droughted, rice rapidly becomes damaged by the effects of low tissue water potential. This is not due to a low capacity to synthesise abscisic acid (ABA), for rice leaves accumulate ABA in response to drought to a greater extent than other cereal species (Henson & Quarrie, 1981). By comparison with paddy rice, upland rice has features – lower tillering and deeper roots – which are associated with drought resistance in other small grain cereals. Evidently, these are not of sufficient importance for its water economy to confer the degree of drought resistance possessed by wheat or barley.

*Salinity*
    The presence of many species of higher plants in coastal salt waters and salt deserts is evidence that they possess features which enable them to survive and grow in saline media. A recent estimate (see Flowers, Haji-bagheri & Clipson, 1986) shows that there are at least 800 species of halophytic angiosperms in more than 250 genera, and that genera with halophytic species comprise 6% of all angiosperm genera. As shown in Table 1, salt marshes and mangrove swamps can be moderately productive. Salt deserts, on the other hand, appear to be much less productive, suggesting either that water is the dominant limiting factor or that plants cannot overcome the combined effects of salinity and drought stress.

    The halophytic habit and mechanisms of salt tolerance have long interested physiologists and have been reviewed on numerous occasions (e.g. by Flowers *et al.*, 1986 and in Staples & Toenniessen, 1984 and Läuchli & Bieleski, 1983). Typically, halophytes grow in media containing 100–250 mol m$^{-3}$ sodium chloride (seawater contains $c$.500 mol m$^{-3}$ NaCl) and sometimes other toxic ions. Considering only NaCl and taking a typical value for the ratio of transpiration to assimilation of 300, and supposing a plant is growing in a medium containing 200 mol m$^{-3}$ NaCl, it can be shown that for each gram of organic dry matter produced, 3.5 g of NaCl have either to be excluded, excreted or sequestered by the plant. Discrimination against Na$^+$ and Cl$^-$ ions is always the first line of defence and sequestering in vacuoles (manifested morphologically as succulence, a high ratio of water to organic dry matter), or in older leaves, is a second line of defence in some plants. Thirdly, some plants possess specialised structures for salt excretion (Fahn, 1979). Halophytic dicotyledons are often succulent and accumulate high concentrations of Na$^+$ and Cl$^-$ ions in vacuolar sap, though the concentrations of these ions in the cytoplasm can also be much greater than in mesophytes. To produce an approximate balance of osmolarity with the vacuole, substantial concentrations of organic solutes such as proline, glycinebetaine, sorbitol or other polyols are present in the cytoplasm, depending on species. Enzymes, membranes and membrane constituents in the cytoplasm need to be tolerant both of the inorganic ions and the organic solutes, or if sensitive, to be isolated from them by compartmentation.

The accumulation of organic solutes, the need to have salt-tolerant enzymes (which may be inherently less efficient than their equivalents in glycophytes) and compartmentation impose an energy cost on halophytes that possess them. In addition, there is likely to be an energy cost associated with discrimination against sodium and the uptake of water against a greater potential gradient than would be encountered by well-watered glycophytes. Thus they may not be as productive as otherwise similar plants which do not possess these halophytic features. An indication of this is that maximum growth rates of halophytes often occur at concentrations lower than those they commonly encounter. Clearly salinity tolerance in dicotyledons must depend on many features, primarily metabolic but also morphological.

Halophytic Gramineae do not have large vacuoles or succulent habit. When grown in saline media they preferentially exclude sodium and, at least in the young leaves, chloride ions. They can also accumulate organic solutes to help to maintain a low osmotic potential. The ability to maintain a high ratio of $K^+$ to $Na^+$ appears to be heritable. Chromosome 4D of hexaploid wheat ($2x = 42$) has been found to influence the $K^+/Na^+$ ratio in leaves. Although there appears to be little allelic variation for this character in hexaploid wheat, variation may exist in *Aegilops squarrosa*, the donor of the D genome, and other relatives of wheat carrying this genome (Gorham *et al.*, 1987). Studies using wheat tetrasomic lines ($2x = 44$) and wheat/*Agropyron junceum* disomic addition lines ($2x = 44$), have shown that chromosomes 2A, 2B and 2D of wheat and 2J of *A. junceum* carry genes which confer susceptibility to salt. However, chromosome 2J also appears to carry genes for salt tolerance (Forster, Gorham & Taeb, 1988). These findings suggest the existence of genes with major effect which might be exploited to increase salt tolerance. At least five different mechanisms determining the $Na^+/K^+$ ratio have been postulated (Jeschke, 1984) so at least five and possibly many more genes may exist for which there could be useful allelic variation (i.e. that conferring a lower $Na^+/K^+$ ratio).

Evidence of the role of organic solutes in maintaining osmolarity and electroneutrality in halophytic Gramineae is equivocal (Wyn Jones, Gorham & McDonnell, 1984). Results of analyses of bulk tissue are difficult to interpret because the location of the organic solutes within cells and within tissues is likely to have an important bearing on their roles. Among different species some solutes may be more important than others, and their relative importance may vary with salinity and with tissue age and condition.

## Prospects
### Drought

Knowledge of the physiology and metabolism of the many species which grow in arid environments is far from comprehensive and there is much scope for further research. It would be surprising if we knew all the mechanisms which plants have evolved to enable them to succeed in such environments. The evaluation of particular characters for which there is genetic variation within a species can be done by conventional genetical techniques coupled with appropriate field experimentation. On the other hand, it is likely to take many years to identify and characterise genes involved in drought tolerance and assess their value in breeding crop varieties with increased drought tolerance. We are only just beginning to learn about the short-term responses to water stress at the molecular level. There appear to be proteins whose synthesis is induced by drought stress (Jacobsen, Hanson & Chandler, 1986; Lalonde & Bewley, 1986; Chapter 9). When the functions of these proteins are discovered it may be deemed worthwhile to manipulate them to enhance drought tolerance.

Genetic variation in drought tolerance within most crop species is probably fairly limited and for two of the world's major crop plants, wheat and maize, has probably been fairly effectively exploited by empirical breeding and selection. As noted earlier, experience indicates that the variation in drought tolerance is likely to be due to the effects and interactions of many genes, most of them with small effects (e.g. Chapter 11; Austin, 1987). The further exploitation of variation in these and other species is likely to be achieved best by conventional breeding programmes involving trials at many locations and over many years. Few of the numerous screening tests which have been proposed for identifying drought tolerance have been adopted by breeders (possible exceptions are described in Chapter 11 and by Blum, 1983 and Clarke & Townley-Smith, 1986). A major reason for this is that most screening tests assess a single attribute (e.g. stomatal conductance) at a single time. Expression of the attribute is likely to vary with the degree of stress, with other features of the environment, and with age of plant and tissue. Further, no single attribute is likely to be the main cause of variation in yield in a drought-prone environment. Finally, to be feasible in a plant breeding programme, screening of large numbers of plants must be possible.

From the above, it can be seen that there is a need for a screening test that is a more specific indicator of drought tolerance than yield itself, that will measure the response of the plant integrated over a substantial part of its life cycle, is non-destructive and can be used for testing large numbers of plants. A test which appears to meet these criteria is carbon isotope discrimination (see Chapter 4). It has long been known that plants discriminate variably against the heavier isotope of carbon (carbon 13) and Farquhar, O'Leary & Berry (1982) have shown that in $C_3$ plants the discrimination is linearly related to the ratio of the intercellular carbon dioxide in leaves to that in the surrounding atmosphere ($c_i/c_a$), which in turn is a measure of the balance between assimilatory capacity and stomatal conductance, and hence water-use efficiency. Thus, where efficient water use is at a premium and leads to increased yield, discrimination would be negatively correlated with yield.

In a series of yield trials with wheat, Condon, Richards & Farquhar (1987) showed that carbon isotope discrimination ($\Delta$) was positively correlated with grain yield. In trials with barley varieties grown in dry Mediterranean environments (Craufurd *et al.*, 1988), it has also been found that the correlation between $\Delta$ and grain yield was positive. In these trials the sampling errors associated with estimating $\Delta$ were lower than those for yield itself, and the correlation of yield with $\Delta$ was stronger than with any of several morphological or developmental attributes measured in the same trials. It remains to be shown why the correlation between yield and $\Delta$ was positive in the barley and wheat trials, and not negative, as a simple interpretation of the theory predicts, but there are several reasonable explanations. Certainly, on the basis of the observed correlations, $\Delta$ is very promising as an indicator of yielding ability in dry environments, but much more work is needed before it can be offered to breeders as a routine screening test.

Transfer of genes between sexually incompatible species is now possible, in principle, using recombinant DNA technology. As noted above, comparative studies of species differing in drought susceptibility may enable targets for gene transfer to be identified. However, it is likely that the best strategy is to exploit species which already have good drought tolerance rather than to attempt to improve the drought tolerance of susceptible species, unless there is a sound reason for so doing. As noted earlier, this is because drought resistance is a complex of characters involving many genes and it is unlikely in the forseeable future that it will be possible to transfer more than one or a few genes from one organism to another by recombinant DNA technology.

Finally, careful management of ecosystems in dry lands is essential to prevent their destruction by inappropriate and uncontrolled farming and grazing. With better knowledge of the biology of the component species and of the ecosystems we will learn how they can be exploited in sustainable ways.

*Salinity*

Some of the earliest experiments on the physiology of plants were concerned with their responses to inorganic ions and the topic rightly remains an important subject of research. Until recently, lack of suitable techniques made it impossible to study the processes at the subcellular and molecular level and so gain an understanding of the mechanisms of selective uptake, compartmentation and transport. Research on this subject is being greatly stimulated by the techniques of chemical microscopy and micro-probing (Läuchli, Spurr & Wittkop, 1970; Van Zyl *et al.*, 1976; Garrison & Winograd, 1982; Ottensmeyer, 1982). Also, as with drought stress, it has been found recently that salinity causes quantitative changes in the relative amounts of proteins synthesised (Hurkman & Tanaka, 1987; cf. Chapter 9). Thus, we can expect to learn much more about the complexes which function as ion transporting and excluding systems. This will apply for essential ions (nitrate, phosphate, potassium, etc.) as well as potentially toxic ones (sodium, chloride, etc.). The techniques will be used to study genetic differences, both within and between species and genera, and we can anticipate being able to identify genes coding for and regulating the functions of the proteins involved in the various aspects of salt tolerance. Genetically defined stocks of species which vary in tolerance will be invaluable in elucidating the mechanisms involved. The knowledge so gained will enable assessments to be made of targets for genetic engineering for increased salt tolerance.

Fortunately, existing knowledge enables saline soils to be managed effectively, albeit at considerable cost. In the long term, however, appro-priate management of salt-affected soils will be necessary to prevent increasing salinisation and degradation of the soils. Thus, with some exceptions, increasing the salt tolerance of crops by conventional or novel means will only 'buy time' by delaying the need to provide the necessary infrastructure to control salinity. Exceptions are in coastal riverine soils in the tropics which are suitable for growing rice, and in areas where there is an abundance of brackish, slightly saline water which can be used for irrigating soils without too high a clay content. In both these cases, the water can be used to prevent progressive salinisation and extreme salt tolerance is not needed. Some progress has been made by conventional breeding and selection of rice and varieties have been produced which yield 3–4 t ha$^{-1}$, as compared with 1–1.5 t ha$^{-1}$ for susceptible varieties (Ponnamperuma, 1984; see also Chapter 12). Among the other grain crops, barley is moderately salt tolerant and considerable genetic variation for tolerance exists in the species and could be exploited more systematically. In the future, it may be possible to transfer genes from salt-tolerant genera such as

*Hordeum*, *Aegilops* and *Agropyron* to wheat (see Chapter 12), and possibly to other cereal species.

In soils too saline for traditional crops, there is the possibility of developing halophytes for crop or forage production. An important consideration is that the produce (e.g. leaves or grain) should contain only low concentrations of sodium or chloride, or it will be unsuitable for human or animal consumption.

### Conclusion

In concluson, few plant physiologists would deny that there remains much to be discovered about the ways in which plants cope with the various kinds of stress they can experience. For crops, man's first line of defence against stress is to avoid it by appropriate management, including choice of crop species. The second line of defence is to breed for increased tolerance to stress. As in other areas of crop improvement, progress has and will continue to be made by empirical selection. A better understanding of the physiological basis of responses to stress and tolerance of stress will contribute to increasing the efficiency of selection for improved performance under stress. Although we cannot estimate with confidence the limits to crop productivity in stressful environments, we can be optimistic that there remains scope for continued improvement.

### References

Austin, R.B. (1987). Some crop characteristics of wheat and their influence on yield and water use. In *Drought Tolerance in Winter Cereals*, ed. J.P. Srivastava, E. Porceddu, E. Acevedo and S. Varma, pp. 321–36. Chichester: John Wiley.

Blum, A. (1983). Genetic and physiological relationships in plant breeding for drought resistance. In *Plant Production and Management Under Drought Conditions*, ed. J.F. Stone and W.O. Wills, pp. 195–205. Amsterdam: Elsevier Science Publishers.

Clarke, J.M. & Townley-Smith, T.F. (1986). Heritability and relationship to yield of excised leaf water retention in durum wheat. *Crop Science*, **26**, 289–92.

Condon, A.G., Richards, R.A. & Farquhar, G.D. (1987). Carbon isotope discrimination is positively correlated with grain yield and dry matter production in field-grown wheat. *Crop Science*, **27**, 996–1001.

Cooper, J.P. (ed.) (1975). *Photosynthesis and Productivity in Different Environments*. London: Cambridge University Press.

Cooper, P.J.M., Gregory, P.J., Tully, D. & Harris, H.C. (1987). Improving water use efficiency of annual crops in the rainfed farming systems of West Asia and North Africa. *Experimental Agriculture*, **23**, 113–58.

Cowan, I.R. & Troughton, J.H. (1971). The relative role of stomata in transpiration and assimilation. *Planta*, **97**, 325–36.

Craufurd, P.Q., Clipson, N.J., Austin, R.B. & Acevedo, E. (1988). An approach to defining an ideotype for barley in low-rainfall Mediterranean environments. In

*Improving Winter Cereals under Temperature and Soil Fertility Stresses*, Proceedings of Cordoba Symposium, 1987 (in press).

Fahn, A. (1979). *Secretory Tissues in Plants*. London: Academic Press.

Farquhar, G.D., O'Leary, M.H. & Berry, J.A. (1982). On the relationship between carbon isotope discrimination and the intercellular carbon dioxide concentration in leaves. *Australian Journal of Plant Physiology*, **9**, 121–37.

Farquhar, G.D., von Caemmerer, S. & Berry, J.A. (1980). A biochemical model of photosynthetic $CO_2$ assimilation in leaves of C3 species. *Planta*, **149**, 78–90.

Flowers, T.J., Hajibagheri, M.A. & Clipson, N.J.W. (1986). Halophytes. *The Quarterly Review of Biology*, **61**, 313–37.

Forster, B.P., Gorham, J. & Taeb, M. (1988). The use of genetic stocks in understanding and improving the salt tolerance of wheat. In *Cereal Breeding Related to Integrated Cereal Production*, ed. M.L. Jorna and L.A.J. Slootmaker. Wageningen: PUDOC. (In press.)

Garrison, B.J. & Winograd, N. (1982). Ion beam spectroscopy of solids and surfaces. *Science*, **216**, 805–11.

Gorham, J., Hardy, C., Wyn-Jones, R.G., Joppa, R.L. & Law, C.N. (1987). Chromosomal location of a K/Na discrimination character in the D genome of wheat. *Theoretical and Applied Genetics*, **74**, 584–88.

Grummet, R., Albrechtsen, R.S. & Hanson, A.D. (1987). Growth and yield of barley isopopulations differing in solute potential. *Crop Science*, **27**, 991–5.

Gutschick, V.P. (1987). *A Functional Biology of Crop Plants*. London: Croom Helm.

Hanson, A.D., Nelson, C.E., Pedersen, A.R. & Everson, E.M. (1979). Capacity for proline accumulation during water stress in barley and its implications for breeding for drought resistance. *Crop Science*, **19**, 489–93.

Henson, I.E. & Quarrie, S.A. (1981). Abscisic acid accumulation in detached cereal leaves in response to water stress. *Zeitschrift für Pflanzenphysiologie*, **101**, 431–8.

Hurkman, W.J. & Tanaka, C.K. (1987). The effects of salt on the pattern of protein synthesis in barley roots. *Plant Physiology*, **83**, 517–24.

Jacobsen, J.V., Hanson, A.D. & Chandler, P.C. (1986). Water stress enhances expression of an $\alpha$-amylase gene in barley leaves. *Plant Physiology*, **80**, 350–9.

Jarvis, P.G. & Mansfield, T.A. (1981). *Stomatal Physiology*. London: Cambridge University Press.

Jeschke, W.D. (1984). $K^+$–$Na^+$ exchange at cellular membranes, intracellular compartmentation of cations, and salt tolerance. In *Salinity Tolerance in Plants*, ed. R.C. Staples and G. Toenniessen, pp. 37–66. New York: John Wiley.

Jones, H.G. (1976). Crop characteristics and the ratio between assimilation and transpiration. *Journal of Applied Ecology*, **13**, 605–22.

Lalonde, L. & Bewley, J.D. (1986). Patterns of protein synthesis during the germination of pea axes and the effects of an interrupting desiccation period. *Planta*, **167**, 504–10.

Lange, O., Nobel, P.S., Osmond, C.B. & Ziegler, H. (1981). *Physiological Plant Ecology II. Encyclopaedia of Plant Physiology*, Vol. 12B, ed. A. Pirson and M.H. Zimmerman. Berlin: Springer-Verlag.

Läuchli, A. & Bieleski, R.L. (1983). *Inorganic plant nutrition. Encyclopaedia of Plant Physiology*, Vol. 15B, ed. A Pirson and M.H. Zimmerman. Berlin: Springer-Verlag.

Läuchli, A., Spurr, A.R. & Wittkop, R.W. (1970). Electron probe analysis of freeze substituted, epoxy resin embedded tissue for ion transport studies in plants. *Planta*, **95**, 341–50.

Mansfield, T.A. & Davies, W.J. (1981). Stomata and stomatal mechanisms. In *The Physiology and Biochemistry of Drought Resistance in Plants*, ed. L.G. Paleg and D. Aspinall, pp. 315–46. Sydney: Academic Press.

Monteith, J.L. (1973). *Principles of Environmental Physics*. London: Edward Arnold.

Morgan, J.M., Hare, R.A. & Fletcher, R.J. (1986). Genetic variation in osmo-regulation in bread and durum wheats and its relationship to grain yield in a range of field environments. *Australian Journal of Agricultural Research*, **37**, 449–57.

Nobel, P.S. (1983). *Biophysical Plant Physiology and Ecology*. San Francisco: W.H. Freeman.

Ottensmeyer, F.P. (1982). Scattered electrons in microscopy and microanalysis. *Science*, **215**, 461–6.

Paleg, L.G. & Aspinall, D. (eds) (1981). *The Physiology and Biochemistry of Drought Resistance in Plants*. Sydney: Academic Press.

Passioura, J.B. (1983). Roots and drought resistance. *Agriculture and Water Management*, **7**, 265–80.

Ponnamperuma, F.N. (1984). Role of cultivar tolerance in increasing rice production on saline lands. In *Salinity Tolerance in Plants: Strategies for Crop Improvement*, ed. R.C. Staples and G.H. Toenniessen, pp. 255–71. New York: John Wiley.

Quarrie, S.A. (1982). Role of abscisic acid in the control of spring wheat growth and development. In *Plant Growth Substances*, ed. P.F. Wareing, pp. 609–19. London: Academic Press.

Salter, P.J. & Goode, J.E. (1967). *Crop Responses to Water at Different Stages of Growth*. Farnham Royal: Commonwealth Agricultural Bureaux.

Staples, R.C. & Toenniessen, G.M. (ed.) (1984). *Salinity Tolerance in Plants: Strategies for Crop Improvement*. New York: John Wiley.

Srivastava, J.P., Porceddu, E., Acevedo, E. & Varma, S. (eds) (1987). *Drought Tolerance in Winter Cereals*. Proceedings of an international workshop, Capri, Italy, 27–31 October 1985. Chichester: John Wiley.

Szarek, S.R. & Ting, I.P. (1975). Photosynthetic efficiency of CAM plants in relation to C3 and C4 plants. In *Environmental and Biological Control of Photosynthesis*, ed. R. Marcelle, pp. 289–97. The Hague: Dr W. Junk.

Turner, N.C. & Kramer, P.J. (eds) (1980). *Adaptation of Plants to Water and High Temperature Stress*. New York: John Wiley.

Van Zyl, J., Forrest, Q.G., Hocking, C. & Pallaghy, C.K. (1976). Freeze-substitution of plant and animal tissue for the localisation of water-soluble compounds by electron probe microanalysis. *Micron*, **7**, 213–24.

Wyn Jones, R.G., Gorham, J. & McDonnell, E. (1984). Organic and inorganic solute contents as selection criteria for salt tolerance in the Triticeae. In *Salinity Tolerance in Plants: Strategies for Crop Improvement*, ed. R.C. Staples and G.H. Toenniessen, pp. 189–203. New York: John Wiley.

Zeiger, E. (1983). The biology of stomatal guard cells. *Annual Review of Plant Physiology*, **34**, 441–75.

# INDEX

regeneration, *see under* protoplast culture,
  tissue culture
regulatory elements, *see* enhancers,
  promoters
rehydration 115, 117–8, 123, 125
relative water content 54, 97, 102
reporter gene 134, 143, 147
reproduction 36, 73, 102
resistance to stress, *see under* stress
  tolerance
resistance to infection 6–8
resistance to water flow 73, 98
resource depletion/limitation 31–2, 35–39,
  43, 102
respiration 55
restriction enzyme 134
restriction fragment length polymorphism
  (RFLP) 142–3, 149
resurrection plants 116–7, 121–2, 124–5
'reverse genetics' 140–3, 145
rheology of cell wall 6, 96, 103–6
ribonuclease 164
ribosomes 163
  *see also* polyribosomes
ribulose bisphosphate carboxylase
  *see* rubisco
rice, *see Oryza sativa*
root 54, 98
  axial resistance 201
  growth 71–89, 103–6, 201–2, 241
  hydraulic conductivity 238
  respiration 55
root:shoot
  growth, *see* shoot:growth
  salt transport 189
  signalling 54, 82–9
root-sourced chemical signal 54, 83–9
rubisco 52, 136–7, 161
Ruderal, *see* strategy theory

saline habitats 99, 102, 237, 241
  soils 217, 219–21, 245
salinity 3, 5–7, 47, 57–8, 144–5, 150–1, 157,
  164–5, 181, 184–92, 236–7, 241–2,
  245–6
  resistance, *see* salt tolerance
  stress proteins 6, 144–8, 161, 164–5
  and yield 218–9, 226–7, 245–6
  *see also* salt
salt
  accumulation in cell wall 110–11
  and enzyme stability 119–20, 123
  excretion 241
  glands 220
  tolerance 95, 108–11, 165, 184–192,
    217–231, 241–2, 245–6
  toxicity 110–11, 241
*Scirpus sylvaticus* 38–9
season

biomass changes with 34–37, 39–42
  growing 35, 207–8, 238
*Sedum telephium* 151
seeds 19–21, 102, 115, 117–8, 121–2, 125
seed coat 172
seedling 3, 77, 208
selection, *see* breeding
selenium 171
senescence 6, 202–6, 221
serine proteinase inhibitor proteins 171
shade
  photoinhibition 62
  plant 31
  reproductive effort 36–8
Sheffield 40
shoot 71–5, 77, 81–9, 118
shoot:root growth 71–89, 201
shoot/root ratio 87
signals 7, 54, 83–9
  *see also* hormonal messages
silver (Ag$^+$) 171
Simon's hypothesis 118
Sitka spruce 13–15
SO$_2$ pollution 8, 21–2
sodium (Na$^+$) 99, 101–2, 109–11, 220, 222–5
sodium chloride, *see* saline habitats, salinity,
  salt
sodium sulphate (Na$_2$SO$_4$) 188
soil
  compaction, *see* impedance, mechanical
  physical properties 39, 42, 47, 54, 57, 220
  toxicity, *see* heavy metals, salinity
soil/root interactions 54, 71–89, 221
soil water potential 2, 48–9, 54, 103, 220
Solanaceae 137, 171
*Solanum* spp. 63, 209, 226
*Solidago* sp. 63
solute
  cell wall 98, 110
  compatible 147
  inorganic 110, 117, 119–20, 182, 184–7
  leakage 118, 121
  organic 73, 77–9, 99, 102, 106, 109, 123–5,
    147, 187–9, 220, 242
  vacuolar 184, 187
  *see also* compartmentation
somaclonal variation 181, 231
somatic hybrid 181, 190
sorbitol 124, 147, 184, 241
*Sorghum* spp. 8, 168, 198, 201–3, 205, 209,
  230
soybean 50, 74, 76, 137, 145, 160–62, 166,
  168–9, 189, 192, 198, 201, 203, 219
*Spartina* sp. 237
spectral properties of leaves 205
spermidine 125
spermine 125
*Spinacia* spp. 52, 60, 219
spores 115, 118